# Adaptive Multimodal Interactive Systems

T0134472

Matthias Bezold • Wolfgang Minker

# Adaptive Multimodal
# Interactive Systems

 Springer

Matthias Bezold
Institute of Information Technology
University of Ulm
Albert-Einstein-Allee 43
89081 Ulm
Germany
matthias.bezold@uni-ulm.de

Wolfgang Minker
Institute of Information Technology
University of Ulm
Albert-Einstein-Allee 43
89081 Ulm
Germany
wolfgang.minker@uni-ulm.de

ISBN 978-1-4899-8600-9          ISBN 978-1-4419-9710-4 (eBook)
DOI 10.1007/978-1-4419-9710-4
Springer New York Dordrecht Heidelberg London

Printed on acid-free paper

Springer is part of Springer Science+Business Media (www.springer.com)

# Preface

Interactive systems, such as personal navigation devices, cell phones, home entertainment and automotive dashboards systems, are part of our everyday lives. Offering more and more features to their users, these systems however increase in their complexity. In addition, the diversity of the users is constantly growing, for instance with older people using interactive systems. One solution for solving the problem of an increasing user diversity and a growing number of features are systems that adapt themselves to individual users. Such an adaptation comprises different steps and relies on an observation of the user's behavior and conclusions drawn from these observations. This process is called user modeling. Adaptations are thereafter selected based on the user modeling information. This book introduces a general framework for adapting multimodal interactive system, comprising a detailed discussion of each of the steps required for such an adaptation.

The observation of user behavior is a prerequisite for performing adaptations. Based on an observation of basic events, such as button presses, speech input, or internal state changes, user preferences are derived. Different algorithms extract information from these basic events, such as preferences of the user or a prediction of the most likely following user action. Additional models support the user modeling process. An interaction model describes user actions. We introduce the use of a task model for describing higher-level user-system interactions and for deriving adaptation triggers, such as predicting user actions and detecting user problems.

In this book, adaptations are presented as a set of adaptation patterns, which are similar to patterns known from software or usability engineering. Patterns describe recurring problems and present proven solutions for these problems. Each of these patterns includes a discussion of the context of use. The patterns provide guidance to the system designer for integrating adaptive features into interactive systems. We introduce a novel set of adaptation patterns. These address both graphical interfaces as well as speech-based and multimodal interactive systems. In addition, we describe an adaptation framework that provides tool-support for creating adaptive interactive systems. For this purpose, the framework introduces an abstraction layer that uses semantic web technology. The adaptations are implemented on top of this abstraction layer by creating an abstract representation of the adaptation patterns, including a system-independent and reusable part and a system-specific part.

In summary, a generic approach for adapting multimodal interactive systems is presented. This approach comprises algorithms for user modeling and a set of adaptation patterns. A reference implementation proves the feasibility of the approach as well as the viability of the user modeling algorithms and adaptation patterns. The evaluation demonstrates that adaptations present a means for improving the usability of interactive systems for an individual user. The conceptual adaptation framework provides a sound foundation for the implementation of adaptations in interactive systems.

# Contents

# 1

## Introduction

The reasonable man adapts himself to the world:
the unreasonable one persists in trying to adapt the world to himself.
Therefore all progress depends on the unreasonable man.

–George Bernard Shaw (1856–1950), Maxims for Revolutionists

John commutes to work every day with his car. When he enters the car, he first selects a radio station in the infotainment system to listen to his favorite morning show. Since other family members also use the same car, John's station is often not selected in the morning. "This is annoying," he thinks. "The car could select the station automatically." Next, he enters the address of his workplace into the navigation system to avoid traffic jams. The system provides a list of favorite destinations, but John still has to enter the navigation menu, open the list of favorites, select an entry, and start the route guidance. "This should be possible with a single action," he mumbles. Sometimes, John uses his wife's car, which has a different dashboard interface. Since John is not familiar with that interface, he would prefer a simplified interface in his wife's car that only offers basic features, whereas he favors the full-fledged interface in his car.

Sometimes, John dreams of entering the car, putting his coffee in the cup holder – and hearing a soft voice: "Good morning, John, I have already tuned in your favorite radio station and entered your work place in the navigation system. No congestion today!" He continues to think: "This could be applied to other devices as well. For instance, the TV system might recommend shows or the video recorder might provide help for programming shows." The vision John is imagining is called adaptation. The adaptation of technical systems to user behavior has been a subject of research for many years. Adaptation starts with an observation of user behavior. For example, the interactive system may recognize John's favorite radio station and observe that John always performs the same actions to program the navigation system. Based on these observations, the adaptive interactive system may provide shortcuts, offer adaptive help, or execute actions automatically.

Today's interactive systems face two major challenges. First, with each new interactive system, new features are being added (cf. Thompson et al., 2005). Second, the diversity of the users increases at the same time, with young and older people using these systems. On the one hand, "digital natives" (see Prensky, 2001) grow up with ever changing interactive systems and do not have problems to adjust to new interfaces. On the other hand, "digital immigrants" are reluctant or unable to embrace new devices. Thus, a single interface cannot support the different requirements and abilities of all users. Adaptation has been recognized as a solution for supporting such requirements that are drifting apart (see Broy, 2006 and Jameson, 2003). In doing so, adaptation may open up an interactive system to a larger user base and therefore increase the success of a product. Adaptation also helps to address further requirements. Laws have been passed that enforce usability and accessibility in interactive systems. For example, the German law of equality for handicapped people ("Behindertengleichstellungsgesetz") from the year 2002 regulates accessibility in government websites. Adaptation may offer specific interfaces for the capabilities of individual people, for instance by changing the font size and colors.

Höök (2000) discusses four challenges that have to be met before adaptive user interfaces become real: usability, useful adaptations, development methods, and maintainability. In this work, we present solutions for these challenges for adaptive interactive systems. We present usability principles and discuss their implications on adaptive interactive systems (*usability*). We devised a set of adaptations and documented them as adaptation patterns, thus sharing best practice of adaptive interactive systems (*useful adaptations*). We created an adaptation framework that allows a developer to integrate adaptive features into interactive systems (*development methods*). We implemented interactive systems and tooling that is available at design time and runtime (*maintainability*). In addition to the challenges presented by Höök, a description of user behavior poses another topic of adaptive interfaces. For this purpose, we also present approaches for user modeling in adaptive interactive systems. Adaptation of interactive systems therefore may be divided into two areas of work: Modeling user behavior and improving the user interface based on these observations. In this work, we present a conceptual framework both for user modeling and adaptations in interactive systems. Thereafter, we introduce an adaptation framework for multimodal interactive systems. A reference implementation of the framework serves as a test bed for an evaluation. In the remainder of this chapter, we introduce the constituents of this work, discuss the general structure of adaptive interactive systems, present adaptation causes other than user behavior, and provide an outline of this work.

## 1.1 Adaptive Multimodal Interactive Systems

Adaptation represents an intriguing idea for improving the usability of interactive systems. Different models of adaptive interactive systems have been investigated in the past decades. This chapter introduces the constituents of adaptive interactive systems and discusses the components common to all adaptive interfaces. Langley (1997) defines adaptive user interfaces as follows:

> An adaptive user interface is an interactive software system that improves its ability to interact with a user based on partial experience with that user. (Langley, 1997, page 56)

This definition establishes two areas of work, user modeling and adaptation. In order to perform adaptations, an adaptive interactive system observes the user-system interaction and represents the user by a model. Such a user model is constructed automatically from an observation of the user or interactively by asking the user during the interaction. The view of the interactive system on the user is limited to the information in the user model. However, this view is always partial, since information beyond the observations is not available to the interactive system. The user model serves as a basis for deciding whether to apply adaptations at all and, if so, which adaptations to select. The purpose of adaptations is to improve the user-system interaction, for example by supporting expert users with repetitive tasks or introducing novice users to the interactive system. In the following, we discuss the main constituents of adaptive interactive systems in more detail, namely interactive systems, user behavior, and adaptations.

### 1.1.1 Multimodal Interactive Systems

Interactive systems refer to computer-based systems with an interface that enables users to interact with these systems. Examples of interactive systems include personal navigation devices in the car, interactive TV systems in the living room, or mobile phones that people carry with them. Interactive systems enable the user to accomplish goals with a device. For this purpose, the user controls functions provided by the device through the interface.

The nature of user-system interactions is manifold. It includes speech interaction, touch screen-based interaction, and input by a remote control. Input and output channels, such as speech or touch-based interaction, are called modalities. Multimodal interactive systems refer to systems with more than one communication channel. Multimodal interaction has found widespread use in the domain of automotive dashboard systems. These systems offer access to functions such as a navigation services, radio, or climate settings by means of a graphical screen, which the driver controls by a rotary push switch or a touch screen. In addition, many dashboard systems support speech-based interaction to allow the driver to keep the hands on the steering wheel while on the road. Speech input and output are used to interact with the system. Gestures, eye gaze, and body movement are further examples of input modalities.

One of the earliest multimodal interactive systems, which is called "put-that-there" (Bolt, 1980), allows a user to interact with a geometrical world by means of spoken deictic commands and gestures, such as "put this cube over there". Twenty years after Bolt, Oviatt (1999) disproves ten myths of multimodal interaction by presenting empirical findings for multimodal interactive systems. For instance, user input from different modalities does not occur concurrently, but with a gap of several seconds. In addition, users do not provide data redundantly, but the inputs from the different modalities complement each other. An example of an interactive system that supports a wide range of modalities, such as speech, gestures, and facial expressions, is the SmartKom prototype (Wahlster, 2003).

In multimodal interactive systems, multimodal fusion combines the signals of the individual modalities. In addition to combining modalities, the fusion extracts semantics and converts the different channels into a common representation. For instance, both speech and haptic input may open the route guidance screen in a navigation device by triggering a specific event. Whereas multimodal fusion is beyond the scope of this work, our approach supports multimodal interaction by considering different modalities in user modeling and adaptation.

### 1.1.2 User Behavior

Adaptations may be triggered by different adaptation causes, such as the context of the interaction, the culture of the user, or user behavior. In this work, we discuss the adaptation of interactive systems to user behavior. From a philosophical point of view, Dretske (1988) defines behavior as an internally produced movement and contrasts behavior with other movements that are produced externally. For instance, raising an arm is behavior, whereas having an arm raised by someone else is not. Reflexive and other involuntary movements are considered as behavior. Thus, behavior does not necessarily have to be voluntary and intentional.

From the perspective of an interactive system, human behavior is limited to actions the system observes, such as clicks of a mouse button, finger presses on a touch screen, or speech utterances. Additional sensors monitor further behavior, such as movements in a room. However, the adaptive interactive system does not differentiate between involuntary and voluntary behavior. Therefore, the system observes user behavior by monitoring input modalities and additional sensors. An adaptive interactive system not only takes into account past user behavior, but also predicts future behavior. This allows the interactive system to anticipate user actions and describe preferences. For this purpose, a user modeling component processes observed user behavior, creates a model from the collected data, and applies user modeling algorithms to this data. Finally, the information in the user model serves as a trigger for adaptations.

### 1.1.3 Adaptations

Adaptation of interactive systems describes changes of the interface that are performed to improve the usability or the user satisfaction. For example, an adaptation may enable a novice user to accomplish an otherwise unsolvable task by offering context-sensitive help or by highlighting interface elements related to the current task. In addition, adaptations decrease the task completion time or improve the satisfaction of the user, without necessarily improving other objective measures.

Adaptations may be performed at different levels of abstraction. At the lowest level, an adaptation changes individual interface elements. For instance, an adaptation may highlight a list entry by changing its color. More complex adaptations add components to the interface or rearrange components, such as graphical buttons or speech output prompts in a voice user interface. Finally, adaptations may alter the dialog flow of the interactive system. For example, an adaptation may select different approaches for specific tasks for beginners and experts. The aim of adaptations is an improvement of the user-system interaction.

## 1.2 Structure of Adaptive Interactive Systems

All adaptive interactive systems share a common structure and include a number of components that are necessary for performing adaptations. In this section, we present these components and discuss their contribution to the process of adapting interactive systems to user behavior. Adaptation to an individual user is either initiated by the user or by the interactive system (Oppermann, 1994; Jameson, 2003). Therefore, it seems to be important to differentiate between adaptable, i.e., manually adapted, and adaptive, i.e., automatically adapting, interactive systems. Adaptable systems are also referred to as customizable or personalizable interactive systems. Most PC software offers settings for customizing software according to individual preferences. For example, the Microsoft Office productivity suite allows the user to arrange the menu items freely. An adaptation component takes the adaptation decision in automatically adaptive interactive systems. For this purpose, the interactive system creates an abstract representation of the user and an adaptation component improves the interactive system to better reflect the user's characteristics. In this book, we focus on interactive systems that automatically adapt to a user.

We now introduce the general structure of adaptive interactive systems and discuss different models that adaptive interfaces employ, such as the user model and the system model. Finally, different adaptation triggers, such as user behavior or context, are introduced.

### 1.2.1 General Structure of Adaptive Interactive Systems

Research has produced a number of different abstract models to describe adaptive interactive systems. These models address the structure of interactive systems as well as the adaptation process. Jameson (2003) regards the user model as the central component of user-adaptive interactive systems. First, information about the user is acquired and stored. Second, the user model is applied to the interactive system and the outcomes of the user modeling process serve as a basis of decision-making for adaptations. Jameson's model however does not include the individual adaptations of a user interface. The ISATINE framework (Lopez-Jaquero et al., 2007) combines the taxonomy of adaptive systems by Dieterich et al. (1993) with Norman's mental theory of action (Norman, 1986). The framework proposes the following stages for adapting interactive systems:

1. Goals for user interface adaptation
2. Initiative for adaptation (user, system or both)
3. Specification of adaptation
4. Application of adaptation
5. Transition with adaptation (between before and after the adaptation)
6. Interpretation of adaptation
7. Evaluation of adaptation

An agent-based implementation of the framework reproduces the individual steps of the framework. However, some of the steps of the ISATINE framework are not necessary for system-initiated adaptive interactive systems, such as "initiative for adaptation".

Brusilovsky et al. (2001) and Paramythis and Weibelzahl (2005) present models of system-initiated adaptive interactive systems. Both models have been created for the purpose of evaluation in the domain of adaptive hypertext systems and share a process-oriented view on adaptation. The process is segmented into the following steps:

1. Monitoring the user-system interaction and collecting input data
2. Assessing or interpreting the input data
3. Modeling the current state of the world (omitted in Brusilovsky's model)

4. Deciding about the adaptation
5. Executing the adaptation

The model presented by Paramythis and Weibelzahl furthermore comprises a number of sub-models. These are divided into static models that do not change at runtime (e.g. system model and action model) and dynamic models, which are updated by the user-system interaction (e.g. user model and interaction history).

Based on these models for adaptive hypertext systems, we created a model for adaptive interactive systems. It is shown in Figure 1.1 and serves as a foundation for the description of adaptive systems in this work. The model builds on the previously presented models and extends them. It describes an iterative process that starts with the user-system interaction and closes with the application of adaptations. On the outside, the processes involved in the adaptation procedure are shown. Inside of the processes, a number of models required for the adaptation are depicted. We separate adaptation into two phases: modeling user behavior and performing adaptations. The first two steps of the model describe user modeling. This phase consists of collecting data by observing the user-system interaction and performing computations with the collected data. For this purpose, a number of user modeling algorithms are presented in Chapter 3. The outcome of user modeling may for instance be a forecast of a future user action, a prediction of a value the user will select, or a statement about the user's proficiency. The final two steps of the adaptation process prepare and execute adaptations. First, once meaningful information has been extracted from the observation of the user-system interaction, the adaptive system employs the user modeling results to decide about adaptations. For this purpose, an adaptation component investigates which adaptations to apply in order to improve the user-system interaction, for instance by supporting the user with repetitive tasks or assisting non-experienced users. The adaptation decision also deliberates whether or not to apply adaptations. Moreover, other currently active adaptations have to be taken into account when deciding about the application of adaptations. Once the adaptation component has identified a possible adaptation and expects an improvement of the user-system interaction from this adaptation, the adaptation is applied to the interactive system. Adaptation is not a one-time change of the interactive system. Instead, the adaptation model represents an iterative process. Once an adaptation was performed, the user continues to work with the interactive system and more interaction data is collected. Based on these new observations, the adaptive system identifies additional adaptations or disables adaptations that are not considered helpful any more. Therefore, we consider adaptation as a continuous process of evaluation and improvement.

The adaptation process relies on a number of models, which are abstract representations of concepts. The view of the system on the respective parts of the system is limited to these models. Figure 1.1 includes four models:

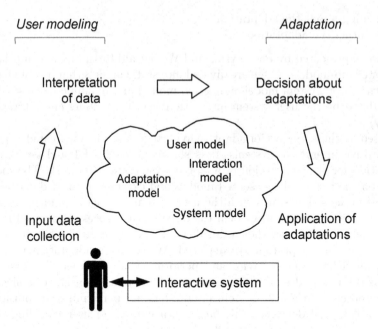

**Fig. 1.1.** A model of adaptive interactive systems.

the system model, the interaction model, the user model, and the adaptation model. In the following, we introduce the individual models and their role in the adaptation process.

### 1.2.2 Models in Adaptive Interactive Systems

In this section, we introduce a number of models that describe different aspects of adaptive interactive systems. These models are, in one form or another, part of most adaptive interactive systems or adaptation frameworks. Some models of other frameworks are not needed when adapting an interactive system to user behavior, such as the context model. Other models found in the literature use different names, but describe the same concepts. For instance, the task model is a separate model in some frameworks (e.g. Paramythis and Weibelzahl, 2005), but part of the interaction model in this work. The individual models used in this work are system model, interaction model, user model, and adaptation model.

#### The System Model

The system model describes different aspects of the interactive system that is the subject of adaptation. This model is mostly static, but reflects changes caused by the adaptations. The dialog flow defines the execution sequence of the application. The reference implementation of the framework presented in this work uses a statechart-based formalism for defining interactive systems. These diagrams consist of states and transitions between these states. Transitions are triggered by events, which are for example caused by user actions.

Different user interface (UI) components define contingent parts of the interface in different modalities, e.g. graphical screens or speech contexts. These components consist of interface elements, such as graphical buttons or speech output prompts. UI components are attached to states and the interactive system activates them when entering the respective state. For instance, the dialog manager displays a graphical screen or plays back the speech output prompts of a speech component. Graphical user interfaces (GUI) interact with the user by means of a graphical display, which depicts one screen at a time. A screen consists of a hierarchy of interface elements. These elements include text labels, buttons, lists, and images. User input is performed by means of touch-based interaction or other control devices, such as a remote control or a push rotary device. Voice user interfaces (VUI) communicate with the user by means of speech output and speech input commands. Loudspeakers play back speech output, which is either pre-recorded or synthesized. The interactive system records utterances by the user with a microphone and performs automatic speech recognition (ASR) to extract the meaning of the utterance. Grammars and statistical language models, which enable natural language understanding, define valid user input. Other modalities, such as gesture input, are defined analogously. All UI components are organized in a hierarchical structure and consist of a main component with a number of children. For instance, a graphical component has a special element type called container that holds other components. Speech components consist of speech input commands and output prompts.

The user-system interaction is bidirectional, i.e., the interactive system communicates with the user and the other way round. User input, such as a button press on the input device or a speech utterance, triggers events that are processed by the dialog manager. For instance, a state change in a state transition diagram initiates the loading of new components. State transition diagrams serve as an example formalism, but others may be integrated as well. Adaptations read information about the interactive system from the system model and write updates to it.

**The Interaction Model**

The interaction model describes actions a user performs as well as the possible relations between these actions. A user action is a logical step and represents the atomic unit of meaningful user interactions. User actions may be described either by a single basic observation or a sequence of these observations. The interaction model comprises a list of user actions that are available in an interactive system. These actions correspond to the functionality of such a system.

An interactive system has to identify user actions at runtime to facilitate a description of user behavior. This information is either emitted by the interactive system, for instance based on annotation at design time, or machine learning techniques identify user actions from an observation of the user-system interaction. The same action may be performed in different modalities and with different input devices, such as speech input, input by haptic devices (e.g. mouse, remote control, or push rotary device), or gesture input. For example, a user may lower the volume of a TV system with a button on the remote control or a speech utterance "lower the volume". Therefore, actions also represent a conceptual abstraction from modalities.

User actions serve as building blocks for a higher-level description of user behavior. Several approaches are available for this description. Plan recognition (Carberry, 2001) refers to the approach of inferring plans from an observation of the user-system interaction. A plan describes possible combinations of user actions that are performed to achieve a goal. Plans often include uncertainty, for instance about the order of actions and the user's goal. Task modeling (Paternò, 2001) presents a similar approach, which explicitly models all possible interactions between the user and the interactive system by means of a task model. Tasks define all actions that may be performed in an interactive system by means of a hierarchical structure composed of based user actions. The higher-level description computes different types of information about the user, such as a list of possible actions or a prediction of the next action. This information finally serves as an adaptation trigger.

**The User Model**

The user model creates a representation of the user that serves as an abstraction of relevant user characteristics. This information in turn triggers adaptations. For this purpose, a user modeling component performs "learning, inference, and decision making" (Jameson, 2003, page 2) and stores the outcome of these computations in a user model. The contents of the user model depend on the specific requirements of an adaptive interactive system. This model comprises information such as user preferences, an interaction history (based on information from the interaction model), goals, or predictions. Kobsa (2001) presents a list of "services" offered by a user model, which includes for example assumptions about user characteristics, a classification of users belonging to specific subgroups, recording of user behavior, and drawing additional inferences. A user model for instance stores information like a classification of the user as beginner or expert or a prediction of the next user action. User models support different data types, ranging from simple flags to complex data structures for sophisticated user modeling algorithms. These algorithms, such as neural networks, Bayesian networks or Markov chains (see Witten and Frank, 2005), derive new information from the data stored in the user model.

The input for a user model is either explicit user input or an implicit observation of the user. The user enters explicit information by means of a questionnaire or a form. These questions include for instance a self-assessment of the user as a beginner or an expert and information about capabilities, such as hearing or sight impairments. An implicit data collection is based on an observation of the user-system interaction. For this purpose, the interactive system collects data from external sensors (e.g. physiological data) or records internal events (e.g. user input or system reactions, such as loading a different screen). The user model is connected to these observations, stores the collected data, and processes it with different user modeling algorithms.

However, the user's characteristics are not constant, but change over time when the user is working with an interactive system. This phenomenon is called concept drift. For example, the user may become an expert user or the preferences of the user change over time. Therefore, user modeling has to be performed perpetually to address these changes. For example, Koychev and Schwab (2000) address drifting interests by means of gradual forgetting.

**The Adaptation Model**

The adaptation model specifies a set of adaptations. In addition, it defines the context in which to apply these adaptations. The adaptation model also comprises a strategy that defines when adaptations should be applied. For instance, an adaptation may improve the interaction for a beginner, but distract or annoy an expert user. The adaptation decision also considers the list of currently active adaptations.

Adaptation descriptions comprise a declarative and a functional part. The former one defines the prerequisites, effects, and improvements of the adaptations. In addition, it connects the adaptation to user modeling information that is used as a trigger. The functional description defines the application of an adaptation to a specific part of the interface. This work includes a set of adaptations that comprise both a declarative and a functional description. For example, the "List Element Selection" adaptation (see Section 4.3.2) highlights interesting items in a list based on a prediction by the user modeling component (declarative description) by changing the background color of the respective item (functional description).

**The Domain Model**

In addition to the previously discussed models, the domain model describes knowledge from the domain of the interactive system. In an interactive TV system, the domain model may for instance describe channels. In a personal navigation device, the domain model may describe cities, streets, and traffic information.

## 1.3 Causes of Adaptation

In this work, we discuss the adaptation of interactive systems to user behavior. However, adaptive interactive systems may also adapt to other causes of adaptation, such as the context of the interaction and the device. When adapting to an individual user, an interactive system may consider different user characteristics. These characteristics include user behavior, invariant characteristics of the user, and emotions. The interactive system observes the user-system interaction to determine user behavior. For this purpose, the interactive system monitors basic events. User behavior includes a list of past interactions or preferences, such as the most frequently selected names in an address book, and predictions about future user behavior.

An interactive system may adapt to other user characteristics than behavior, such as age and hearing or vision impairments. However, these kinds of characteristics cannot be extracted from the user-system interaction and require a manual personalization of the adaptive system by means of explicit user modeling. For instance, Schneider et al. (2007) describe a decrease of the performance of simple sensory, motor and cognitive activities after the age of 65 and propose adaptivity as a solution. Different kinds of motor impairments hamper the operation of regular input devices, such as a mouse. Some impaired users have problems to click, whereas others have difficulties in moving the mouse. Gajos et al. (2008) present an adaptive user interface that limits the required number of mouse clicks or distance a mouse has to be moved according to individual capabilities. User characteristics may be represented in a user model as static facts and the adaptations are executed in a similar fashion as in interactive systems that adapt to user behavior. Another facet of the user is the culture, which has implications for the interaction style and appearance of the interactive system. Reinecke and Bernstein (2009) present an interactive system that adapts to the different countries a user has lived in.

Emotions represent another user characteristic that an adaptive interactive system may consider. The current emotional state of the user, which may change quickly, is extracted from speech (Kwon et al., 2003) or from facial expressions (Cohen et al., 2003). Negative emotions, such as anger or frustration, are more relevant for adaptive interactive systems, since positive emotions indicate that the dialog is working well. If the user is angry, the interactive system may adapt the strategy. For instance, the ultimate adaptation in a telephony system is to switch from an automatic dialog to a human operator. Pittermann et al. (2007) and Gnjatović and Rösner (2008) present examples of emotion-aware adaptive dialog systems.

User interfaces do not only adapt to user characteristics, but also to other factors, such as the context of use or the device. Context-aware user interfaces (Harter et al., 1999) consider the current context of use. For instance, a context-aware museum guide (Chou et al., 2005) not only considers the knowledge of the user, but also incorporates the currently visited room.

Different devices have significantly different capabilities, such as screen sizes or computing resources. A device-adaptive application automatically exploits the resources of a device optimally. The diversity of screen sizes is most notable with mobile devices, ranging from small mobile phones to tablet devices. For example, Eisenstein et al. (2001) present a model-based framework for developing mobile applications that adapt to the characteristics of a specific device. Thus, a wide range of adaptation causes exists. Our work however addresses the adaptation of interactive systems to user behavior. Since the user model represents the adaptation cause, it may be extended to support other causes as well.

## 1.4 Outline

This work is structured as follows. In Chapter 2, we present on overview of related work. First, we address different adaptive interactive systems, namely adaptive hypertext systems, adaptive user interfaces, and adaptive speech dialog systems. Thereafter, different using modeling algorithms and architectures are presented. We introduce the concept of design patterns and adaptation patterns as well as the formalization of patterns. Finally, we discuss how interactive systems employ semantic technologies for dialog management and in adaptation architectures.

In Chapter 3, we introduce an approach for modeling user behavior from basic events. We regard user behavior as a combination of actions and data. An approach for recognizing user actions with probabilistic automata is presented. Based on these actions, a description of higher-level user actions becomes feasible. Task models provide a description of user actions. We derive information such as a list of possible actions or detecting user problems from these models. In addition, a prediction of user actions triggers different adaptations. We present two algorithms for predicting one action and an algorithm for predicting a sequence of actions.

In Chapter 4, we present a number of usability principles for interactive systems and discuss their implications for adaptive interfaces. We introduce design patterns as a method for sharing best practice in a domain. In order to define adaptation patterns for interactive systems, we created a specific pattern format. Thereafter, we present a set of multimodal adaptation patterns for interactive systems. These patterns address both graphical and speech user interfaces.

In order to support the implementation of adaptations in interactive systems, we present an adaptation architecture in Chapter 5. This framework creates an abstraction of the interactive system by means of semantic technologies. It includes a user modeling component and an adaptation component. A reference implementation shows the practicability of the framework and serves as a test bed for an evaluation. In Chapter 6, we present an evaluation of the adaptation approach. We tested the action prediction algorithms with recorded user sessions. In order to investigate the adaptations, we created different test systems and performed user tests with these systems. For the evaluation, objective measures were collected by means of log files and subjective measures were collected with a questionnaire. In Chapter 7, we summarize this work and present future research.

# 2

# Related Work

All intelligent thoughts have already been thought;
what is necessary is only to try to think them again.

–Johann Wolfgang von Goethe (1749–1832)

Adaptive user interfaces and adaptive dialog systems have been a topic of research for decades and a significant number of prototypes and approaches have been developed. In this chapter, we present adaptive systems and adaptation approaches and discuss their relation to this work. Adaptation may be regarded as the application of artificial intelligence techniques to user interfaces. The aim of artificial intelligence is to create intelligent computer systems by imitating – and possibly surpassing – human intelligence. In doing so, artificial intelligence enables a computer to become a worthy opponent in playing games such as chess, to understand human speech, or to become an intelligent agent for tasks like planning appointments. When applied to user interfaces, artificial intelligence techniques are used to observe the user-system interaction and to perform improvements of the interface. Interactive systems that use artificial intelligence to anticipate user behavior and improve the user interface are called intelligent or adaptive interfaces. Adaptations enable a user to work with interactive systems more easily and conveniently.

Research on the adaptation of interactive systems to abilities, traits, and preferences of individual users started with in investigation of user modeling. For instance, user modeling determines user preferences or assigns users to different groups. One example of an early user modeling system is the Grundy system (Rich, 1979), which describes the user by means of stereotypes. Based on these stereotypes, novels are recommended to the user. Subsequently, user modeling was applied to enable the adaptation of the interface to an individual user. One of the earliest investigations of the viability of adaptive interfaces is presented by Greenberg and Witten (1985). A user test with an adaptive menu-driven telephone book showed promising results and led to the development of further adaptive interactive systems. In the following decades, adaptivity has been applied to different kinds of interactive systems, such as office suites, hypertext systems, or speech dialog systems.

In this work, we address the adaptation of multimodal interactive systems in general. We present approaches for user modeling and adaptation that are applicable to a wide range of interactive systems. However, some of the adaptive interactive systems discussed in literature deal with very specific aspects of user interfaces or employ application-specific user modeling algorithms. Examples of such interfaces are an adaptive navigation engine (Bachfischer et al., 2007) or an adaptive restaurant guide (Langley, 1999). The user modeling algorithms and adaptations used in these interactive systems are not applicable to user interfaces in other domains. Instead, we discuss generic approaches that may be applied to different interactive systems. Adaptive systems also include recommender systems, for instance for online shops or movies based on films the user has watched before. An overview of recommender systems is given by Adomavicius and Tuzhilin (2005). A special case of recommender systems are collaborative filtering systems (Breese et al., 1998), which generate recommendations from preferences of other users. However, recommender systems are not the focus of this work. In the following, we provide an overview of different research areas in the domain of adaptive interactive systems and introduce examples.

This chapter is structured as follows. First, we give an overview of adaptivity in different kinds of interactive systems. Next, we describe different approaches for user modeling and introduce the concept of design patterns. Finally, a review of the use of ontologies and semantic technologies in interactive systems concludes the chapter.

## 2.1 Adaptive Interactive Systems

Adaptivity has been applied to a wide range of interactive systems, such as adaptive hypertext systems, adaptive user interfaces, or adaptive speech dialog systems. This section gives an overview of these different types of adaptive interactive systems.

## 2.1.1 Adaptive Hypertext and Adaptive Websites

The adaptation of hypermedia systems and websites has been investigated extensively. Hypermedia systems provide interconnected content nodes to the user and the user moves between these nodes by means of hyperlinks. The content is composed of text fragments, graphics, and movies. Examples of hypermedia systems include learning systems, museum guides, or websites on the internet. Adaptive hypermedia systems generate custom versions of hypermedia pages for an individual user, for instance by adjusting content nodes to the knowledge level of the user or by sorting links according to the user's interests. A user modeling component constructs a user model from an observation of visited nodes and time spent at each node. Based on this information, the user model identifies other content nodes the user might be interested in, for instance by recommending unknown nodes for relevant topics. Thereafter, different adaptation techniques generate a custom presentation for the user. Adaptive websites, a special case of adaptive hypertext systems, apply adaptation techniques to websites (Perkowitz and Etzioni, 1997). User modeling in adaptive websites is usually based on mining log files produced by a web server and employs algorithms such as cluster mining (Perkowitz and Etzioni, 1999). Since hypermedia systems are focused on content, user modeling algorithms primarily are concerned with the documents a user accesses and the adaptations alter the presentations of the content. The user interaction in interactive systems is richer and allows extracting more extensive information. For instance, hypermedia systems are usually limited to log data, whereas interactive systems observe the interaction of the user with each interface element, such as scrolling in a list or speech interaction.

Adaptations in adaptive hypermedia systems have been categorized into two groups, adaptive presentation techniques and adaptive navigation support (cf. De Bra et al., 1999a). Adaptive presentation performs adaptations of content nodes, such as showing or hiding nodes or selecting among different versions of a node. For instance, longer or additional text fragments are selected for starters, whereas only a short text is presented to experts. In addition, the adaptation may gray out known content fragments. Adaptive navigation support adapts the link structure between content nodes to reflect the knowledge and preferences of a user. For instance, links between content nodes are emphasized by reordering them to show important ones on the top or by annotating links with text or graphics. Links to nodes that are not interesting for the user are hidden by showing them as regular text or removed by not showing the link text at all. A comprehensive discussion of adaptive hypermedia technologies is presented by Brusilovsky (2001). It includes the taxonomy of adaptive hypertext technology given in Figure 2.1. The taxonomy includes adaptive presentation techniques and adaptive navigation support and divides these technologies into more specific adaptations. However, adaptations for hypertext systems may not be transferred directly to interactive systems. Adaptations can be applied to a wide range of different

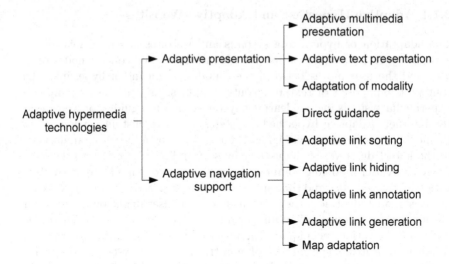

**Fig. 2.1.** A taxonomy of adaptive hypermedia techniques (based on Brusilovsky, 2001, page 100). The lowest layers have been omitted for reasons of clarity.

interface elements rather than content nodes and links. User interfaces consist of components, such as buttons or speech input and output elements, which require different adaptations. Therefore, we present specific adaptations for interactive systems in this book. However, the advent of client-side dynamic web pages, which enable a web page to communicate with the server and update the page dynamically, blurs the differences between user interfaces and hypermedia systems. For instance, Schmidt et al. (2008) present an adaptation architecture for dynamic semantic web pages.

Different reference models have been defined in the domain of hypermedia, such as the Dexter reference model (Halasz and Schwartz, 1994). A reference model is an abstract definition of a domain and the concepts therein. It facilitates a general discussion of a domain, without restricting the validity to a single system. Different approaches extend the Dexter reference model for adaptive hypertext systems. The Adaptive Hypermedia Application Model (AHAM; De Bra et al., 1999b) adds three sub-models to the storage layer of the Dexter reference model. First, the domain model describes the content of the hypermedia system and comprises a list of text and multimedia fragments and links between these nodes. Second, the user model stores information about the knowledge level of different users, such as a list of visited content fragments. Third, the teaching model consists of a set of "pedagogical rules" and defines how adaptations are performed. The Munich reference model (Koch and Wirsing, 2002) is similar to AHAM, but uses the Unified Modeling Language (UML) as a formal foundation. This model also extends the storage layer of the Dexter reference model and introduces a domain model, a user model, and an adaptation model, which correspond to the respective models in AHAM. Based on these models, different operations, such as authoring, retrieval, or adaptation, specify the functionality of adaptive hypermedia systems. Other domains of adaptive systems may benefit from this research, which has produced sophisticated models for adaptive hypertext systems. In this work, we use an ontology for information representation, which may serve as a foundation for a reference model.

In the following, we present examples of adaptive hypermedia systems. ELM-ART (Weber and Brusilovsky, 2001) is an intelligent tutoring system for the LISP programming language. A traffic light metaphor provides adaptive navigation support: a green bullet in front of a link recommends a page for a user, whereas a red bullet indicates that the user's level of knowledge is deemed insufficient for that page. The AVANTI project (Fink et al., 1996; Stephanidis et al., 1997) provides multimodal access to a tourist information system and considers the different needs of individual users. For instance, the AVANTI system provides laypersons with an explanation of technical terms, which is not necessary for expert users. One special focus group of the system are handicapped and elderly people. For instance, the tourist information contains more information on accessibility of sites when used by wheelchair users. A specific version of the text is served to blind people on a Braille keyboard. The AVANTI system supports adaptability, i.e., a manual adaptation by the user, as well as adaptivity, i.e., an automatic adaptation by the system. Adaptations are performed by means of adaptation rules.

Adaptive hypermedia techniques may also be applied to the semantic web, which extends regular hypertext with semantic annotations. These encode the meaning of the respective text sections and facilitate information extraction and reasoning. For example, the annotations clearly identify names of people or time designations in hypertext documents. Adaptive hypertext techniques have been applied to semantic websites. For instance, Dolog et al. (2003) present SIMPLE, an adaptation framework that is based on semantic web technologies and rule-based reasoning. For this purpose, a rule-based language called TRIPLE, which is presented in more detail in Section 2.4, defines adaptation rules. The approach presented in this work also relies on semantic web technology. However, instead of rules, we use a more abstract representation of adaptations. At the same time, the adaptations may be reused and tailored to the requirements of a specific interactive system.

### 2.1.2 Adaptive Graphical Interfaces

Graphical interfaces enable users to control a wide range of devices, such as personal computers, mobile phones, or personal navigation devices. Due to the high degree of deployment, significant research on adaptive user interfaces has been conducted in the domain of office suites, such as Microsoft Office. Since a lot may be learned from an investigation of unsuccessful approaches, we start this section with a discussion of unsuccessful adaptations and expose their shortcomings. Thereafter, we introduce those adaptive interfaces that have proven successful.

One well-known use of adaptive technology is the Office Assistant, a cartoon style animated help agent introduced in Microsoft Office 2000 and removed again in Microsoft Office 2007. The agent observes user actions and provides help to the user. The technology is based on the Lumière research project (Horvitz et al., 1998), which employs explicitly modeled Bayesian networks to infer interesting help topics. However, different reasons caused the help agent to fail, such as a high degree of distraction and the selection of irrelevant help topics. Swartz (2003) presents a discussion on peoples' attitudes towards anthropomorphic interface agents and identifies some of the problems related to the Office Assistant. For example, agents are not able to comply with human rules of etiquette. In general, interface agents are active and personalized collaborators that assist the user in an intelligent way with various tasks. Maes (1994) presents interface agents, which support the user with tasks such as handling incoming mail in an e-mail application, planning appointments in a calendar application, or filtering news in a news reader. However, Shneiderman (see Shneiderman and Maes, 1997) argues that direct manipulation interfaces correspond to how humans perceive and think better than intelligent agents. Direct manipulation describes interfaces in which elements on the screen are directly manipulated, e.g. with a mouse. The use of visualization techniques allows the user to deal with huge amounts of data.

Adaptive menus (see Figure 2.2) are a feature that was added to Microsoft Office and removed later. The menu hides rarely used menu entries and the user expands the full menu by clicking on small arrows at the bottom of the menu. However, if an item is not in the short menu, the user has to read both the short and the full menu, making the selection of items that are not on the short list very time-consuming. In addition, this kind of adaptive menu breaks with the usability principles of learnability, predictability and consistency by not leaving items in a place in which users always find them. Mitchell and Shneiderman (1989) obtained similar results in a comparison between static menus and dynamic menus. Dynamic menus placed frequently used items to the top of the list. An evaluation showed that dynamic menus did not improve the interaction and users preferred static menus. Therefore, if elements are removed from a list, the list element is more difficult to find, for instance because the user cannot remember positions in a list. These examples show a reluctance of users to adopt adaptive interfaces that do not adhere to usability principles. However, other research projects demonstrate that adaptive interfaces improve both performance and user satisfaction. In this work, we discuss usability principles for adaptive systems and present adaptations that comply with them.

Extensive research has been performed on adaptive menu selection. Two kinds of adaptations for list items were identified: changing the position by moving or duplicating menu items (spatial), and changing the appearance of menu items (graphical). One successful spatial adaptation of menus are split menus, which place frequently used items into a separate section on top of the list. For example, Microsoft Word uses split menus in the font selection box. Frequently used fonts are shown in a separate section on top of the list. A study by Sears and Shneiderman (1994) shows that split menus, which move elements into the split area instead of copying them (see Figure2.3(a)), accelerate list selections and users prefer split menus over non-adaptive menus. Findlater and McGrenere (2004) compared a static split menu, an automatically adaptive split menu, and an adaptable split menu, which lets users put items into the top section by hand. An evaluation revealed that the adaptable version was faster than the adaptive version and users preferred the former one. This suggests that users prefer customizable menus to adaptive menus. However, a study by Mackay (1991) showed that users do not perform manual customization unless the advantage of doing so is obvious. Moreover, many input devices limit the possibilities of complex customization, such as small mobile phone keypads or push rotary switches in the car. Gajos et al. (2006) investigated different adaptive versions of toolbars, which are common elements of graphical interfaces. For this purpose, three different adaptations were compared to a non-adaptive baseline. First, a split interface puts frequently used items into a separate toolbar. This corresponds to a split menu, which copies elements instead of moving them (see Figure 2.3(b)). Second, a moving interface moves frequently used elements from a popup menu into a list. Third, a visual pop-out interface highlights frequently used elements. An evaluation revealed that the split interface performed best, both yielding an improvement of the user performance and achieving the best ratings in the user satisfaction. The findings of Gajos et al. reveal that duplicating items rather than moving them improves the user performance.

Graphical adaptive menus sustain spatial consistency and only alter the appearance of menu entries. An adaptive emphasis is for instance accomplished by highlighting items that were predicted by a user modeling component. Figure 2.3(c) presents an example of an adaptation that highlights

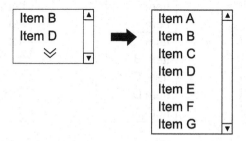

**Fig. 2.2.** Adaptive menu with manual expansion. Only frequently used items are visible when the menu opens and the expands the full menu by clicking on the arrows at the bottom.

menu entries. Tsandilas and schraefel (2005) present an evaluation that compares a list with highlighting to a list with highlighting and shrinking of text. Selection times are lower for the adaptive version with a constant font size and the authors reason based on a worst-case estimate that the adaptive versions should be faster than a non-adaptive baseline. Another comparison of different adaptive menus is presented by Park et al. (2007). They compared a traditional menu to an adaptable menu, which lets users change the sort order of the list, an adaptive split menu, and an adaptive menu that highlights the most frequently selected elements. Again, adaptable menus were most efficient and the users preferred them. While the adaptive highlight menu did not reduce the selection time, users preferred it to the traditional menu. Instead of highlighting menu entries, ephemeral adaptation (Findlater et al., 2009; see Figure 2.4) reduces the visibility of items that are not selected frequently and fades these items in quickly less than one second after the menu appeared. In doing so, it includes the temporal dimension. An evaluation revealed that menus with ephemeral adaptation reduce the selection time and users prefer them to other versions. Menu selection is an important task for any kind of interactive system. Therefore, the adaptations presented in this section are incorporated into a set of patterns we introduce in Section 4.3. Two of the systems created for the evaluation of the patterns use menu selection as their task.

Mixed initiative adaptation (Horvitz, 1999) represents a compromise between user-initiated customization and system-initiated adaptation. This approach reflects the user's preferences by letting the user decide when to employ adaptations. Users prefer being in control, but do not personalize the system unless they see a clear advantage in doing so (cf. Mackay, 1991). MICA (Bunt et al., 2007) is a mixed-initiative adaptation framework that employs online GOMS analysis. This framework recommends frequently used interface elements and the user may place them into a personalized interface. The user

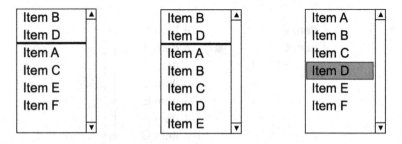

(a) Split menu, items are moved.    (b) Split menu, items are copied.    (c) Menu with highlighted item.

Fig. 2.3. Three kinds of adaptive menus.

decides if the recommended changes should be performed. A user evaluation showed that users prefer the mixed-initiative to a customizable interface, as long as the user modeling is accurate.

The adaptations presented in this section so far alter an existing interface. However, some adaptive systems generate the user interface according to the preferences or capabilities of an individual user. The SUPPLE system (Weld et al., 2003) adapts graphical interfaces to different requirements, for instance for desktop computers and mobile phones or pointer-based and touch panel-based devices (Gajos and Weld, 2004). In addition, a storyboard example is presented in which a printing dialog is adapted by adding frequently used options from sub-dialogs to the main dialog. An extended version, called SUPPLE++, supports users with motor impairments (Gajos et al., 2008). Some users have difficulties in moving the pointing device, whereas others have problems to click. Therefore, SUPPLE++ generates an interface that is tailored to the capabilities of an individual user by reducing either the distance the pointing device has to be moved or the number of clicks. An evaluation showed that the users strongly prefer the adaptive version. The most serious issue of generated interfaces is however the aesthetical appearance of the interface, which is less appealing than an interface designed by a human. In addition, safety requirements cannot be ensured with generated interfaces. Our approach does not employ generated interfaces, but improves interfaces created by a human designer. However, the "Alternative Elements" adaptation presented in Section 4.3.3 allows an interactive system to select among different alternatives provided by the developer.

### 2.1.3 Adaptive Speech Interfaces

Two human partners in a speech dialog adjust to each other, for instance by asking clarifying questions or raising their voices when they think the dialog partner does not understand them because their voice is too soft. Therefore, a

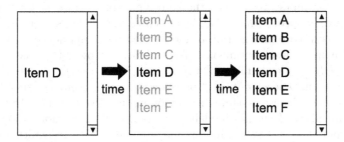

**Fig. 2.4.** Adaptive menu with ephemeral adaptation. Only the item that should be recommended to the user is visible at the beginning. Other items appear quickly thereafter.

natural and intuitive dialog between a machine and a person needs to be able to adapt to the user, rather than having a fixed dialog script. Consequently, adaptation in speech dialog systems is an important research topic. Different methods exist for implementing dialog control in speech-based interfaces. In this section, we introduce these methods, explain how adaptation is implemented within the individual approaches, and discuss the relation of these concepts to this work.

A first kind of dialog specification is based on a rigid dialog definition, which is followed closely by a dialog manager, such as state-based and form-based dialog systems (cf. McTear, 2004). State-based dialog systems define the dialog logic by means of state transition networks. These consist of states and transitions. Speech output by the dialog system is connected to states. Actions, such as speech input by the user, trigger transitions to other states. In doing so, the network defines all possible dialog paths. One example of an adaptive state-based dialog system is the TOOT system (Litman and Pan, 2002), a train information system that automatically adapts itself to the current user. For this purpose, it changes the dialog initiative and the confirmation style. The dialog system starts with a "user initiative", in which the system asks open questions and leaves the initiative to the user, and "no confirmation" strategy, which does not confirm user input. Once problems are detected based on low speech recognition scores, the TOOT system switches to more conservative dialog strategies, such as "system initiative", in which the dialog manager asks the user specific questions, and "explicit confirmation", in which the dialog system confirms user input before advancing to the next question. User evaluations showed that an adaptive version of the TOOT system achieved a higher task success rate than a non-adaptive version. Hassel and Hagen (2006) follow a similar approach and present an adaptive voice-controlled infotainment system deployed in an automobile dashboard system. Speech output prompts are adapted to the expertise level of the current user. For instance, novice users receive a list of available commands, whereas the dialog system only plays short feedback tone for expert users. Table 2.1 presents an example of speech interactions with a novice and an expert user. In general, expert prompts are shorter than prompts for beginners. Rather than considering only speech input confidence scores, the dialog system by Hassel and Hagen uses more information to model the user's experience, such as the number of help requests or timeouts. An evaluation showed that the adaptive prototype reduces the number of turns and the interaction time. The approach of selecting speech output prompts and input grammars depending on characteristics of the user-system interaction matches well with the framework presented in this work. A user modeling component derives information about the user from the user-system interaction, such as speech recognition errors. The adaptations presented in Chapter 4 include adaptations that select the appropriate speech input and output components for a user. Whereas state transition networks are rigid compared to other approaches, they are well suited for defining multimodal interactive systems and facilitate a uniform

| Novice | Expert |
|---|---|
| user: Entertainment. | user: Entertainment. |
| system: Entertainment. You can say AM, FM, or CD. | system: Entertainment. |
| user: Choose CD. | user: Choose CD. |
| system: Say a CD number. | system: Number? |
| user: *Unintelligible.* | user: *Unintelligible.* |
| system: I could not understand you, repeat. | system: Pardon me? |

**Table 2.1.** An example of a speech interaction that is adapted to the experience of the user. (Source: Hassel and Hagen, 2006, page 2.)

definition of different modalities, such as speech and graphics. The reference implementation presented in this book employs a statechart formalism. It allows the development of graphical and speech-based interfaces with a common specification.

Form-based dialog systems also employ a rigid formalism. Forms define a number of slots that have to be filled by the user. The system directs the dialog, enabling the user to provide data for each of these slots. Form-based dialog systems offer more flexibility by letting the user decide whether to fill all slots with a single utterance or one slot at a time. Adaptation in form-based dialog systems is achieved by selecting among predefined components for speech input and speech output. Veldhuijzen van Zanten (1998) presents an example of an adaptive form-based dialog system, which extends the form-filling paradigm with hierarchical slots. Different kinds of questions are defined, such as high-level and low-level or open and closed questions. If speech recognition problems are discovered, the system moves to lower levels in the slot hierarchy, for instance by asking for values of individual slots instead of open questions. If the user provides more data in an utterance than is anticipated by the current hierarchy level (over-information), a higher level in the hierarchy is selected. The user model (Veldhuijzen van Zanten, 1999) consists of a set of flags for each slot that indicate the knowledge level of the user. VoiceXML[1] is a form-based standard for defining voice interfaces. Niklfeld et al. (2001) present an adaptive architecture for multimodal dialog systems that is based on VoiceXML. Whereas the framework implemented within this work does not support the form-based approach, it may be integrated in addition to state-based dialogs. For this purpose, forms have to be attached to specific states in the dialog system.

Other approaches for the specification of dialog systems provide a higher degree of flexibility, for example the Information State Update (ISU; Larsson and Traum, 2000) approach. An information state "represents the information necessary to distinguish it from other dialogs" (Larsson and Traum, 2000, page 1) and a set of dialog moves trigger updates to the information state,

---
[1] VoiceXML: http://www.w3.org/Voice/

for instance by means of update rules. In the TRINDI toolkit, a dialog move engine computes the next action based on the current information state. In order to enable adaptivity in the ISU non-adaptive approach, the TALK project proposes an extension of an ISU-based dialog manager (Georgila and Lemon, 2004; Lemon et al., 2006) with reinforcement learning techniques to enable the system to learn an optimized dialog strategy. Since the ISU approach is a formalism for speech-based dialog systems and does not support multimodal interaction, our framework does not support this approach.

Statistical dialog systems regard dialog design as an optimization problem by handling the dialog flow as a sequential decision process. For this purpose, statistical algorithms, such as Markov decision processes (MDP), learn optimal dialog strategies. Levin et al. (2000) and Scheffler and Young (2002) first estimate the parameters of a simulated user from a dialog corpus and then employ reinforcement learning to find the optimal dialog strategy based on interactions of the simulated user with the dialog system. Different techniques exist for collecting the dialog corpus of training sessions, for example Wizard-of-Oz experiments (Rieser and Lemon, 2008) or training sessions with a preliminary dialog system (e.g. in the NJFun dialog system, Singh et al., 2002). An evaluation showed that the trained dialog strategy outperforms fixed strategies proposed in the literature. All possible states, i.e., all possible values of the variables, form a state space and the dialog strategy defines which action (e.g. which question to ask) should be taken in the current state. A as a result, the dialog moves on to another state. The result of the MDP training is a single optimized, yet non-adaptive dialog strategy. When applying the learning algorithm at runtime, the dialog system adapts to an individual user. For example, the CLASSiC project (Janarthanam and Lemon, 2008; Rieser and Lemon, 2009) employs online learning of a statistical dialog system. Partially observable MDPs (POMDP) extend MDPs in a way that they can handle both unobservable states and uncertainty (e.g. the user's beliefs) and enable the dialog manager to track all possible dialog paths rather than just the most likely path. An application of POMDPs in dialog systems was shown to create dialog strategies that perform better than the ones created by regular MDPs (Young et al., 2010). Paek (2006) presents a discussion of statistical dialog systems. The main advantage of statistical dialog systems is the theoretical foundation, which most other dialog specifications lack. In addition, the statistical approach handles uncertainty well. On the other hand, Paek mentions the reluctance of application developers to give up control over their application as the main disadvantage. Moreover, if the results of statistical dialog systems are not always superior to handcrafted systems, the handcrafted approach is likely to be followed. In addition, statistical approaches are not suited for graphical or multimodal interactive systems and thus aggravate the development of multimodal interactive systems by impeding a uniform approach. Since this book addresses multimodal interactive systems, we do not include statistical dialog systems in the framework.

Agent-based software architectures employ autonomous software components, called agents, to solve complex problems through collaboration. Each of these intelligent agents implements a strategy or competence and contributes autonomously to solving specific tasks. The Interact project (Jokinen et al., 2002) is an example of agent-based dialog systems. It employs an architecture called Jaspis. The system consists of managers, agents, and evaluators. Managers take care of specific components, for instance a dialog manager or a presentation manager. Agents handle (possibly very specific) situations, such as speech recognition errors. Different agents implement different strategies for the same situation and evaluators select among the available agents. The advantages of the agent-based approach are a high degree of flexibility and the possibility to implement different strategies and select among them at runtime. The Jaspis architecture is cross-platform and distributed. However, agent-based architectures bring along a software overhead, which is not feasible with the limited computing resources of many interactive systems. Moreover, the behavior of agent-based systems is less predictable than the behavior of more rigidly defined dialog systems.

In addition to user behavior and traits like expertise, the modality of speech offers emotion as an additional channel. If a dialog system recognizes emotional cues, such as anger or impatience, in the user's voice, the dialog strategy may be adapted accordingly. Examples of such systems are NIMITEK (Gnjatović and Rösner, 2008), which adapts spoken help messages to the current emotional state of the user, and PROBLEMO (Pittermann et al., 2007), an emotion-aware intelligent architecture for dialog systems. Emotions may be integrated into this framework by loading the results from the emotion recognition algorithms into the user model and using them as adaptation triggers. However, we focus on user behavior rather than other adaptation triggers in this book. Thus, we have presented different approaches for adapting speech dialog systems. The approaches that support multimodal interaction integrate well with our framework, whereas other may be added separately from other modalities. We present a set of adaptation patterns in Section 4.3 that include a discussion of speech dialog systems.

## 2.2 User Modeling for Adaptive Interactive Systems

An adaptive interactive system observes the user to find characteristics and preferences in the user-system interaction. These characteristics trigger adaptations, such as shortcuts for repeated actions or adaptive help. The process of observing a user and drawing conclusions is called user modeling. Adaptive interactive systems represent the user by means of a user model. In this section, we give an overview of user modeling approaches in the literature.

The user model obtains information about a user either explicitly by asking the user questions or implicitly by observing the user without interference ("loophole observation"). User-supplied information is for instance collected by means of on-screen dialogs or questionnaires. However, user-supplied information is not reliable and users may be reluctant to provide information about them. In the remainder of this section, we address automatic user modeling that derives information from an observation of the user-system interaction. The information that a user model stores depends on the requirements of the adaptive interactive system and includes goals, knowledge, interests, traits, experience, and preferences.

One of the earliest user modeling systems is the Grundy system (Rich, 1979), which offers novel recommendations to a user. The system regards a user model as a "collection of good guesses about the user" (Rich, 1983, page 200) and includes information such as age and level of education. Based on this information, the current user is assigned to a predefined stereotype. Many user modeling approaches deal with specific domains and problems, such as modeling a user's favorite TV program to be able to recommend interesting shows (e.g. Ardissono et al., 2004 or Bachfischer et al., 2007). In this section, we instead discuss reusable and generic user modeling algorithms and architectures.

### 2.2.1 User Modeling Algorithms

User modeling algorithms serve different purposes in user modeling systems, such as predicting user actions or modeling user preferences. The selection of these algorithms depends on the requirements of the interactive system and the adaptations. This section presents different kinds of algorithms that are suitable for adaptive interactive systems.

Sequence prediction algorithms (SPA) predict a future item based on past items. For instance, SPAs enable an adaptive interface to anticipate a user action based on previously observed actions and offer support accordingly. Davison and Hirsh (1998) present an SPA called Incremental Probabilistic Action Modeling (IPAM), which employs first-order Markov chains and reduces the influence of older data. In an evaluation with UNIX command line actions, a prediction accuracy of 40 % was achieved. However, unlike many interactive systems, UNIX commands do not have a context, i.e., the user may enter all commands at all times. Hartmann and Schreiber (2007) present an SPA algorithm called FxL and compare it to different other SPAs, including IPAM. Different test sets are used, including the UNIX test set by Davison and Hirsh and an office suite test set. Prediction accuracy was limited to a range of 40 % to 60 %, but Hartmann and Schreiber reason that domain knowledge could improve the prediction accuracy. In this work, we present a sequence prediction algorithm for user actions with a similar background as IPAM. However, we employ domain knowledge in the form of a task model to filter out predictions that are not valid in the current context. For this purpose, we adapted a Markov chain-based algorithm for predicting link in websites presented by Sarukkai (2000) to user action prediction. We combined it with a task model to add domain information for better prediction results.

In doing so, the algorithm is optimized for the use in adaptive interactive systems.

Whereas SPAs predict a single item, sequence mining algorithms identify frequently occurring sequences. Adaptive interfaces use sequence mining for example for offering shortcuts for frequent action sequences. Mannila et al. (1997) present a sequence mining algorithm that retrieves frequent episodes from a sequence of events, for instance in user log files. However, the algorithm does not make a statement about the meaning of the discovered episodes. Liu et al. (2003) present an application of Mannila's algorithm to adaptive user interfaces. The algorithm automatically detects repeated action sequences in a word processing software, such as repeatedly applying a certain combination of formatting options, and offers shortcuts to the user for this formatting. In Section 3.4.2, we present an adaptation of Mannila's algorithm for predicting action sequences in adaptive interactive systems. This algorithm was used for the evaluation of an adaptive system (see Section 6.4.6).

In addition to user action prediction, the user's goals and needs serve as adaptation triggers. One approach for modeling goals and needs are Bayesian networks, a graph-based model for probabilistic relationships between random variables. Horvitz et al. (1998) present an agent-based adaptive help system called Lumière that employs Bayesian networks to describe the user's experience and derive help messages for the current situation. Bayesian networks are explicit representations of a domain and a user's knowledge. They facilitate probabilistic inference for determining the user's needs and goals. Dynamic Bayesian networks allow incorporating temporal aspects by integrating the user's interaction history. However, Bayesian networks have to be modeling largely by hand and thus add a considerable development effort in more complex domains. Bayesian networks may be integrated into the user model of our framework.

Different machine learning algorithms (see Witten and Frank, 2005) may be used for user modeling. Algorithms may be divided into supervised or unsupervised ones. Webb et al. (2001) discuss general requirements of machine learning algorithms for user modeling. For instance, these algorithms require a large collection of training data and labeled datasets that have been annotated before. Supervised machine learning algorithms, such as Markov models or neural networks (Mitchell, 1997), train models from labeled data and employ these models to classify unknown individuals by assigning them to a known class. In adaptive interactive systems, classification is for instance used for identifying users or assigning them to a group, such as beginner or expert. For example, Galassi et al. (2005) present an approach that uses Hierarchical Hidden Markov Models (HHMM) to train user profiles from recorded sessions and employ these models to identify users. Unsupervised machine learning algorithms do not rely on labeled data. Clustering algorithms divide a group of elements into a set of clusters. However, the properties shared by different elements in a group are not known. Clustering algorithms have for instance been used in adaptive web systems (Hamilton et al., 2001) to compute page recommendations for new users based on interaction patterns of other users. However, this approach does not work for interactive systems, because interaction data from other users is usually not available. Zukerman and Albrecht (2001) provide an overview of predictive statistical models, such as Markov models and Bayesian networks. These algorithms may be implemented within the user modeling framework presented in this work. The selection of the algorithms depends on the requirements of a specific interactive system. We present a number of user modeling algorithms that we developed or adapted for adaptive interactive systems in Chapter 3.

### 2.2.2 Plan Recognition and Task Models

Plan recognition is the process of observing user actions and determining the goal a user tries to accomplish. A plan represents the order of actions required to reach that goal. Plans include uncertainty, since the actual behavior of users is to some degree unpredictable. Moreover, information such as the planner's intent cannot be observed and may therefore only be inferred with a certain probability. Plans are very complex, especially when they describe "real world" problems, such as cooking. Many plan recognition systems rely on a plan library, which is created by hand with a significant effort. An early example of plan recognition systems called BELIEVER is presented by Schmidt et al. (1978). Carberry (2001) describes a number of problems that need to be addressed in conjunction with plan recognition. First, the input data is often noisy, i.e., the individual observations that are used to infer the user's goal are uncertain. Second, the plan recognizer has to decide among competing hypotheses, e.g. when certain observations are part of different plans. Moreover, users tend to work on different tasks in parallel. Third, plan recognition algorithms have to scale up in large domains with plan libraries that contain numerous plans a user possibly works on.

In this work, we introduce an approach for adaptive interactive systems that employs a technique called task modeling (Paternò, 2001). It describes the user-system interaction by means of tasks, which define possible user actions without speculation about the user's intentions. Task modeling is used in software engineering for design and evaluation. A detailed introduction to task modeling is given in Section 3.3. Instead, we employ task models at runtime to derive information about the user-system interaction, such as predicting future actions or identifying situations in which a user requires assistance. We present task models as a viable means for describing higher-level user behavior in adaptive interactive systems and deriving adaptation triggers from it.

Klug and Kangasharju (2005) employ task knowledge for supporting a user. Their interactive system observes the user's activity by instantiating a task model at runtime and generates an improved user interface to better support the current task. In a similar approach, an intelligent classroom (Franklin et al., 2002) recognizes user actions by means of a plan-based action description and supports users in performing these actions, e.g. by advancing slides during presentations. Mitrović et al. (2008) present an agent-based adaptive user interface that employs a task model notation based on ConcurTaskTrees (Paternò et al., 1997), similar to the approach presented in Section 3.3. Our approach uses the task model for instance for predicting user actions and detecting user problems.

### 2.2.3 User Modeling Architectures

In addition to the algorithms, the architecture of a user modeling system is an important aspect of the user modeling process. In early adaptive interfaces, user modeling used to be an integral component of these systems and could not be reused for other applications. Generic user models (Kobsa, 2001) instead allow the developer to reuse the user modeling architecture. In addition, generic user modeling systems may enable different applications to share a single user model and thus facilitate a reuse of user data. One example of a generalized user modeling system is the Doppelgänger user modeling system (Orwant, 1995). This system supports learning by means of different techniques, such as linear prediction and Markov models. However, custom learning techniques cannot be added and the accumulation of data is left to the application. The user modeling system developer for this work was created such that it allows the integration of a wide range of algorithms. A user modeling framework for intelligent learning environments that implements both supervised and unsupervised learning is presented by Amershi and Conati (2007). More recent developments in the domain of user modeling are ubiquitous and distributed user modeling systems (Heckmann, 2005). They combine information from different sources, such as mobile phones or portable computers. In addition, location awareness is an important topic for mobile devices (Sharifi et al., 2004). Our adaptation framework comprises a generic and extensible user modeling component. The framework may be extended with arbitrary user modeling algorithms and connected directly to observations from the user-system interaction. In doing so, the framework meets the requirements of different kinds of adaptive interactive systems.

Logic-based systems represent data in a way that allows inference of new knowledge by means of logical reasoning. One example of a user modeling system that employs a logic-based representation is the BGP-MS system (Kobsa and Pohl, 1995). It converts the internal representation to first-order logic and uses logic reasoning on the data. Rules draw additional inferences about the user. In a fashion similar to logic-based architectures, semantic user modeling systems employ a semantic representation using semantic web technology. The semantic web employs a description logic formalism to add semantic annotations to existing data, thus allowing a description of meaning as well as reasoning. An ontology defines the structure of the semantic data and often serves as a foundation for semantic user modeling architectures. Various ontologies have been proposed specifically for user modeling. For instance, Golemati et al. (2007) and Heckmann et al. (2005) demonstrate sophisticated examples of such ontologies and store user modeling data in terms of ontologies. The semantic notation facilitates data sharing between different applications. Razmerita et al. (2003) present a general architecture for a user modeling server that is based on semantic technologies. However, semantic user models do not support complex derivations and specific data types (e.g. matrices for Markov chains). Since many algorithms employ such data types, we present a generic user modeling system that supports arbitrary types efficiently. In order to connect the user model to a semantic description of the interactive system, we use a bridging component.

## 2.3 Design Patterns for Adaptations

Every domain has experts who have collected valuable experience about successful solutions for recurring problems. Sharing expert knowledge helps other people working in the same domain to straightforwardly select suitable solutions. Design patterns are an approach to record experience by writing down recurring problems in a specific domain and successful solutions for these problems. In doing so, patterns serve as a means for communicating best practice. Design patterns have their origins in architecture. In the seminal 1977 book "A Pattern Language", Alexander et al. (1977) introduce a set of 253 architectural patterns. Each of these patterns describes an architectural problem and presents a proven solution to this problem. Patterns are defined as follows:

> *Each pattern describes a problem which occurs over and over again in our environment, and then describes the core of the solution to that problem, in such a way that you can use this solution a million times over, without ever doing it the same way twice.* (Alexander et al., 1977, page x)

All these patterns employ the same structure, which is often referred to as the Alexandrian form. The description is narrative and uses few subheadings. After the heading, the context of the pattern is introduced followed by a problem statement and an elaborate discussion of the problem. Thereafter, a concise solution statement and a detailed discussion of the solution, including a small drawing, present the solution. For example, the "Your Own Home" pattern (Alexander et al., 1977, page 392) states that people are not happy in a home they do not own. The recommended solution is to enable people to own the house they live in, which they arrange according to their wishes and expectations.

The concept of design patterns was adopted for the domain of software engineering and design patterns have become an established method in software engineering. The most well-known set of design patterns for software engineering is presented in the book "Design Patterns: Elements of Reusable Object-Oriented Software" by Gamma et al. (1995), often referred to as the "gang of four" patterns. These design patterns describe problems occurring frequently during (object-oriented) software development and present successful solutions. For instance, the "Singleton" design pattern proposes a method to ensure that only a single instance of a specific class exists, which different parts of a software application share. Design patterns have been widely adopted and are part of many students' curricula. Another pattern collection called "Pattern-Oriented Software Architecture" (POSA; Buschmann et al., 1996) presents a set of patterns for software architecture, which deal with a more high-level view on software design. Beyond these patterns, numerous pattern collections have been presented at workshops like "Pattern Languages of Programming" (PLoP), such as patterns for parallel programming (Ortega-Arjona, 2009) or patterns for creativity (Georgiakakis and Retalis, 2009).

A pattern language is a collection of patterns that describe best practice for a specific domain. The patterns cover the most common problems of this domain. An interlinking further connects the patterns to each other. While many patterns do not intersect, some patterns may address overlapping problems. In this case, the context of the pattern determines which pattern to use best. In the following, we introduce the concept of patterns and review patterns from the literature, both in the area of interface patterns and formalization of design patterns.

### 2.3.1 Interface and Adaptation Patterns

Whereas design patterns are mostly used in the domain of software engineering, patterns have also been written for the domain of user interface design and interaction design. Borchers (2001) introduces a pattern language for interactive exhibits. For instance, the "Attract Visitor" pattern emphasizes the importance of having an exhibit that is interesting enough to attract a visitor's attention. Van Duyne et al. (2006) present a pattern language for websites. These patterns are structured into groups such as genres of sites, content, or navigation. For example, the "Sitemap" pattern proposes a single place called sitemap that may be used to access all pages of a website.

Different pattern catalogs have been defined for user interface design. Van Welie and van der Veer (2003) have compiled extensive pattern collections of reusable interface design knowledge, which may be used by designers and developers for creating graphical interfaces. Similarly, Tidwell (2005) presents an extensive structured catalog of 94 interface design patterns. These patterns cover a wide range of topics, from the general structure of graphical applications to form input to aesthetics. For example, the "Extras On Demand" pattern advises a designer to offer a limited set of options in an interface and to add a button that opens a larger set of options for advanced users. The "Cancelability" pattern recommends offering a way to the user for cancelling time-consuming operations. Many interface designs pattern are very specific and describe a single interface element. While some of the patterns describe dynamic behavior, such as "Responsive Enabling" or "Smart Menu Items", these patterns do not address adaptive user interfaces. Dearden and Finlay

(2006) present a detailed and critical discussion on the current state of pattern languages in human-computer interaction.

Patterns have also been used to describe adaptative hypertext systems. A basic set of abstract adaptation patterns was presented for adaptive hypertext systems by Danculovic et al. (2001), introducing "Link Personalization", "Content Personalization", "Structure Personalization", and "Remote Personalization". For example, the "Link Personalization" pattern recommends altering the link structure of a website to better reflect a user's needs. The "Content Personalization" pattern discusses offering personalized content nodes. The patterns are more general in nature than interface design patterns and the number of patterns is significantly lower. These adaptation patterns were extended by Koch and Rossi (2002) with more detailed patterns, such as "Adaptive Anchor Annotation" or "Adaptive Sorting of Anchors". The "Adaptive Anchor Annotation" pattern discusses how adding annotations to hypertext links lets the user better estimate the usefulness of a link. Another pattern called "Adaptive Sorting of Anchors" recommends sorting anchors in a way that more interesting anchors are shown first. As discussed in Section 2.1, adaptive hypertext systems differ from other interactive systems. For instance, the adaptations alter interface elements rather than links between nodes or text. Therefore, we present a distinct set of adaptation patterns for interactive systems in Chapter 4.

### 2.3.2 Formalization of Design Patterns

Design patterns use a textual and narrative form. For example, the patterns collections by Gamma et al. (1995) and Buschmann et al. (1996) have been published as books to be read by programmers. Therefore, humans read the pattern descriptions, but a computer cannot process them automatically. Approaches for the formalization of patterns aim to increase the utility of patterns by representing patterns using a well-defined structure and vocabulary. In doing so, they provide a standardized and machine-processable representation. The formalization of patterns takes place on different levels of abstraction. First, a formalized pattern format extends the narrative description with special markings to label sections of a pattern description. For instance, all patterns may share the "motivation" and "solution" sections. This machine-readable structure ensures consistency and enables referencing between different pattern collections. The Pattern Language Markup Language (PLML; Fincher et al., 2003) follows this approach and provides an XML document type definition (DTD) for specifying patterns. PLML is however a very high-level definition for describing textual pattern collections in a uniform way to enable interchange.

A more formal notation of patterns may serve as a basis for intelligent tool support. For instance, tools may provide support for refactoring existing projects according to patterns (e.g. Zannier and Maurer, 2003). Other approaches employ formal languages to specify patterns, including the semantics of the pattern. Mikkonen (1998) discusses an approach for formalizing patterns based on a custom notation for defining objects formally, with the focus being on the temporal behavior of design patterns. Hallstrom and Soundarajan (2008) present an approach for ensuring implementation correctness and facilitating reasoning about patterns. The approach is based on a pattern contract formalism with pre and post conditions. However, approaches that rely on formal languages have not found wide-spread use due to their high complexity (Zdun, 2007).

Ontologies are formal representations of a domain and the contained concepts. The Web Ontology Language (OWL; Smith et al., 2004) is a language for defining ontologies. Different approaches use OWL to define design patterns. For instance, Dietrich and Elgar (2005) present a formal pattern description based on OWL that describes the structure of design patterns. The formalized patterns are used for scanning source code for design pattern usage and thus assist in the task of documenting source code. Henninger and Ashokkumar (2006) propose a meta-model for software patterns based on an OWL infrastructure for applying patterns in the software development process. This model conceptually builds on PLML, but extends it considerably by including a description logic representation of the patterns. For this purpose, pattern attributes, such as "hasProblem", "hasContext", or "hasSolution", are defined by means of OWL restrictions. Henninger presents interface patterns as an example of this approach. A tool called BORE supports interface development by offering context-sensitive information. Moreover, a formal framework for creating an interconnected pattern language for interactive systems is provided. Our approach provides tool support for adaptation patterns and includes a semantic description of parts of the patterns based on OWL. However, it does not fully formalize the semantics of patterns in OWL. Instead, OWL serves as a common formalization of different aspects of adaptive interactive systems and enables the use of semantic techniques, such as reasoning.

Model-based development (Schmidt, 2006) is based on an abstract model (called platform-independent model or PIM), which is transformed into a concrete model (called platform-specific model or PSM). For instance, Petrasch (2007) presents how to apply formalized interface patterns to model-based user interface development. The more specific patterns are the better they may be incorporated into the model-based development process. We present an integration of a set of adaptation patterns into a model-based adaptation framework. The model-based system description is automatically transferred to an OWL-based semantic layer.

## 2.4 Semantics in Interactive Systems

Semantic interactive systems rely on an ontology for a description of the domain and other topics, such as the interactive system or the user. Ontologies create a formalization of a domain and describe all concepts contained in this domain. For instance, the domain of cooking consists of ingredients and recipes. Both concepts (e.g. vegetables as a group of ingredients) and instances (e.g. a tomato as a special kind of ingredient) form the ontology. The ontology is either created manually, for instance by using semantic editors such as Protégé-OWL (Knublauch et al., 2004), or by mining data automatically, for instance by means of ontology mining (e.g. Buitelaar and Ramaka, 2005). The Web Ontology Language (OWL; Smith et al., 2004) represents a common notation for the definition of ontologies. A more detailed introduction to OWL is given in Chapter 5.

In this section, we discuss the use of semantic technologies in interactive systems. On the one hand, domain knowledge encoded in ontologies allows reasoning on this data, for instance in speech-based dialog systems. On the other hand, ontologies enrich the dialog logic of interactive systems. After a review of the use of ontologies in different research projects, we investigate different ontology-based adaptation architectures.

### 2.4.1 Ontologies in Dialog Systems

A formalized representation of the domain of an interactive system facilitates different applications in dialog systems. On the one hand, a generic dialog engine accesses the domain knowledge when processing user input. On the other hand, a single ontology serves as a unified knowledge representation for different components of an interactive system. Milward and Beveridge (2003) present an approach for replacing hand-crafted dialog design with a generic dialog engine and ontological domain knowledge, which is used for different purposes. For instance, the order in which the dialog system asks the user questions is improved by incorporating the ontology. Other applications of this ontology are in speech recognition and interpretation. In addition, they present a system that supports general practitioners in the decision whether a patient should be referred to a cancer specialist. For this purpose, the system integrates medical domain knowledge encoded in an ontology.

Due to the broad area of application, ontologies have been used in multimodal systems for other purposes than dialog management. In the following, various uses of ontologies in three research projects are presented. The SmartKom project (Wahlster, 2003) created a "dialog shell" for applications that employ multimodal interaction between users and interactive systems, supporting modalities such as speech and gesture for input and speech and graphical for output. The SmartKom system employs an extensive unified ontology (Gurevych et al., 2003), which is utilized by different components of the system. Prior systems instead often used different knowledge bases for different purposes. Porzel et al. (2003) present several applications of the ontology, such as multimodal fusion, semantic coherence scoring (interpretation of the ASR results), and computing dialog coherence (interpretation in context). In addition, the ontology is used for dialog management. For this purpose, a plan language models the actions necessary to carry out processes that are defined by the ontology. The SmartWeb project (Wahlster, 2007) builds on the SmartKom project and provides multimodal access to the semantic web. As an example application, access from a mobile device to information in the domain of a soccer championship is presented. The SmartWeb Integrated Ontology (SWIntO; Sonntag et al., 2007 and Oberle et al., 2007) forms the foundation of the dialog system. Factual knowledge is directly encoded into the ontology and may be queried by the user. Moreover, a discourse ontology specifies different kinds of multimodal interactions. In addition, semantic web services are connected to the ontology (e.g. asking for available activities). The different components of the dialog manager communicate via ontology instances. Thus, the SmartWeb project uses the ontology as a unified knowledge representation for different tasks.

The TALK research project addresses adaptive multimodal interactive systems and includes different components. First, the SAMMIE dialog manager (Becker et al., 2006), an in-car dialog system for an MP3 application, uses an ontology to model both domain knowledge as well as possible tasks. The tasks are automatically converted into a format that a plan-based discourse manager processes. A different approach was followed in the intelligent home scenario (Amores et al., 2006). A domain ontology is used by a knowledge manager component in a dialog system to provide domain reasoning. For instance, in a home automation setting for handicapped people, a query like "are there red lamps in the house" is answered using a query to the knowledge manager, which computes the requested information and offers the results to the user. The approach presented in this work also employs a single ontology, which comprises information about the interactive system, the user, and the domain. However, the adaptations are defined explicitly and are not derived by means of ontology reasoning.

### 2.4.2 Architectures for Adaptive Interactive Systems

In addition to serving as a unified knowledge representation, ontologies have been used for modeling interactive systems. For instance, Obrenović et al. (2003) employ an ontology in addition to UML in the development process of multimodal interactive systems. The high-level model description is only available at design time to provide development support and for platform mapping. However, it is not available at runtime as a knowledge representation for the interactive system itself. Our framework instead uses the model both at design time and runtime. Aragones et al. (2007) present a semantic adaptation framework, called ACUITy. A controller component mediates between a UI engine and an ontology, which comprises information on the user, the user interface, and the domain. An interface is generated from an ontology-based definition of the application. Custom-tailored interfaces are created by incorporating information about past interactions.

Sophisticated adaptation architectures have been presented in the domain of adaptive hypertext systems. Dolog and Nejdl (2003) and Henze et al. (2004) present an approach that transfers techniques from adaptive hypermedia to the semantic web. A first-order logic language adds semantic information to documents and resources, such as topics of a document and dependencies between concepts. For instance, concept B discussed in document X may require prior knowledge of concept A. A rule language called TRIPLE implements reasoning rules, for instance for deriving a set of examples for a specific topic. Since the user modeling is also defined by means of semantic triples, reasoning may be used to extract a custom set of documents or resources for a user, for instance a set of documents that discusses knowledge required as a prerequisite for the current document.

In a similar fashion, the ODAS domain ontology presented by Tran et al. (2008) is used in conjunction with adaptation rules in an adaptive hypertext portal. Tran et al. reason that rules ensure a better transparency and controllability for developers than statistical adaptation methods. The ontology provides a knowledge foundation for adaptation rules and contains different models, such as a system model, a task model, and a resource model. Carmagnola et al. (2005) present a different semantic adaptation framework for hypertext systems. It relies on different planes for the ontological representation a specific type of knowledge, such as the user, the user's actions, the domain, or the context. Rules in the SWRL[2] notation define intersections of the ontological planes, since rules combine the information from these planes. In doing so, rules implement both user modeling and adaptations. User modeling rules infer information about the user and adaptation rules perform adaptations, such as removing links or adding explanations to links. For example, an adaptation rule may emphasize items for older users ("user model" plane) in the night ("context" plane) on a PDA device ("device" plane), thus exploiting information from different planes. However, these architectures do not represent adaptations in an intuitive and reusable way. Writing semantic rules requires special knowledge of semantic technologies. The adaptation framework presented in this work also relies on a unified semantic information representation, on top of which adaptations are defined. Adaptations are triggered by information from the user model, such as a prediction of the next user action. However, adaptations are defined by means of adaptation patterns to facilitate reuse. The ontology does not cover every aspect of the system, but only information required for deciding which adaptations to apply.

## 2.5 Discussion

In this chapter, we reviewed related work for adaptive interactive systems and user modeling. The chapter started with a review of different kinds of adaptive

---

[2] SWRL: A Semantic Web Rule Language Combining OWL and RuleML: http://www.w3.org/Submission/SWRL/

interactive systems, such as adaptive websites, graphical user interfaces, and speech dialog systems. Next, user modeling was introduced as a prerequisite for adaptations. Both different user modeling algorithms, such as machine learning or task modeling, and user modeling architectures were presented. Thereafter, the concept of design patterns was introduced and different design pattern collections for graphical user interfaces and adaptive hypertext systems were presented. Different approaches for formalizing patterns were introduced. Finally, the use of ontologies in interactive systems and semantic adaptation architectures was presented.

However, when addressing adaptive interactive systems, such as digital TV systems or automotive dashboard systems, the approaches presented in the literature cannot be directly transferred, since adaptive interactive systems have different requirements with regard to user modeling and adaptations. For example, hypertext systems adapt documents rather than interface elements. In addition, many approach in the literature present approaches that are limited to specific domains or types of interactive systems. Therefore, we present an approach for user modeling in adaptive interactive systems in the following chapters and define adaptations for interactive systems in a general and reusable way by means of multimodal adaptation patterns. In addition, we present a generic adaptation framework. The framework employs semantic technologies, but offers a reusable and abstract adaptation definition.

# 3

---

# User Modeling in Interactive Systems

I can calculate the motion of heavenly bodies,
but not the madness of people.

–Sir Isaac Newton (1642–1727)

Adaptive interactive systems observe the user-system interaction and draw conclusions from these observations about user characteristics and preferences. The process of representing a user by means of an abstract model is called user modeling. Based on these conclusions, adaptations improve the interactive system. Thus, user modeling represents a prerequisite for the adaptation of interactive systems. The adaptation to user behavior is the focus of this book. In this chapter, we present domain-independent algorithms we developed for modeling user actions. Numerous algorithms and architectures for user modeling have been presented in the literature (see Section 2.2). However, these are often domain-specific, intended for other systems than interactive systems, such as hypertext systems, or describe only a part of the user modeling process. An integrated user modeling approach for adaptive interactive systems needs to extract information from basic observations and process this data to facilitate a higher-level description of user behavior. Finally, this information provides triggers to an adaptation component. These triggers include the recognition and prediction actions or the detection of problems. The approaches presented in the literature do not fulfill these requirements.

In this chapter, we introduce a novel approach for describing user actions in interactive systems. The approach starts with an observation of basic events that occur in the user-system interaction. Different algorithms build on each other to enable a higher-level description of user behavior and to build a user model. User actions are described by sequences of basic events and probabilistic automata detect these actions. Task models in turn facilitate a description of more complex user behavior. Finally, different algorithms derive information from the user model that triggers adaptations. For this purpose, we present algorithms for predicting actions and action sequences.

An example illustrates the different approaches and algorithms we present in this chapter. A digital TV system allows a user to watch different shows and browse an electronic program guide (EGP). Since the number of shows is extensive, the user may narrow down the list by selecting filter criteria, such as channel or time. The TV system provides a graphical interface and may be controlled by means of a remote control and speech input. For example, the user presses the red button on the remote control to open the EPG and uses speech input to select the channel "BBC" as a filter criterion. We use this example throughout this chapter to illustrate various algorithms.

The chapter is structured as follows. First, the notions of user behavior as a sequence of basic events and user behavior as actions and data are introduced. Next, an approach for recognizing user actions in a sequence of basic events is presented. Thereafter, the concept of task models is introduced and the application of task models in adaptive interactive systems is discussed. Finally, we present approaches for predicting a single user action and a sequence of user actions that we devised and adapted for interactive systems.

## 3.1 User Behavior in Interactive Systems

In this section, we introduce the notion of user behavior as a sequence of basic observations or low-level events. We regard user behavior in interactive systems as a combination of user actions and associated data.

### 3.1.1 User Behavior as a Sequence of Events

For a human expert who observes the user-system interaction, the meaning of the interaction is obvious: if a user presses the red button on a remote control, this action tells the interactive system to open the EPG. However, the system has to rely on sensors and internal information to understand the user behavior. For instance, the system may derive from an observation of a specific button press and reactions of a dialog component that a graphical screen called "EpgView" was opened. An understanding of these observations allows a description of higher-level user behavior. The source of these observations depends on the type of interactive system. For example, websites provide log files with page access information and online stores accumulate articles that customers bought. Based on this information, an adaptive website or store may recommend pages or articles respectively to the user. Log events include events from user input, such as a remote control or speech utterances, and the corresponding internal reactions of the system.

Interactive systems have to represent observations of the user in a common format in order to enable a user modeling system to draw further conclusions. We call these observations basic events. They include user input, such as key presses and speech utterances, or system reactions, e.g. internal state or property changes. The view of an interactive system on user behavior is limited to these events. Therefore, we regard the user-system interaction as a sequence of low-level events (cf. Dix et al., 1993). An example sequence of events from the digital TV system presented in the introduction is given in Figure 3.1. The user presses a button "COMM_0x6e" (line 1) and the system responds with a sequence of reactions (lines 2–4). Thereafter, the user selects a channel with a speech utterance (line 5) and the system updates the graphical screen accordingly (line 7).

The source of basic events is not limited to the interactive system, but external sensors may contribute to the stream of events as well, including for instance physiological observations like pulse or the location in a room. Since the view of the user modeling component is limited to basic events, data that should be incorporated into the user modeling process has to be represented by these events. Without loss of generality, the observation of user behavior is therefore limited to the observation of basic events.

Log data in adaptive interactive systems is either processed online or recorded for offline processing. In order to include characteristics of the current user, a user modeling component applies user modeling algorithms at runtime. In some cases, recorded log data is preferable, for instance for training an initial user model or for static user characteristics that do not change over time. For this purpose, the interactive system writes observed events to log files and the user modeling system processes them offline. Moreover, the event types that contribute to the user modeling may be limited to events that are relevant for the adaptations. For example, internal state changes that describe the inner workings of the interactive system instead of user actions might not be relevant for the user modeling.

In order to describe higher-level user behavior, the interactive system has to determine the meaning of sequences of basic events. For instance, a specific sequence of these events may describe one single user action. The higher-level interaction is composed of user actions, which in turn are sequences of basic events. Therefore, recognizing these actions plays an important role in user modeling. However, basic events do not directly reveal which action a user performs. For example, line 1 of Figure 3.1 shows that a button "COMM_0x6e" was pressed. This event does not reveal the meaning of this input. In this case, this action opens the EPG. In addition, a speech utterance, such as

```
1 [1180520776220] hw name={COMM_0x6e}
2 [1180520776220] event name={OpenEpg}
3 [1180520776220] state name={EpgMain}
4 [1180520776376] view name={EpgView}
 . . .
5 [1180520778231] asr value={channel b-b-c}
6 [1180520778242] command name={SelectChannel} value={BBC}
7 [1180520778247] viewupdate name={EpgView}
```

**Fig. 3.1.** Exemplary log lines of a digital TV system. The user presses a button (line 1) and the system loads the EPG screen (lines 2–4). Thereafter, the user selects channel "BBC" with speech input (line 5) and the system updates the screen (lines 6–7).

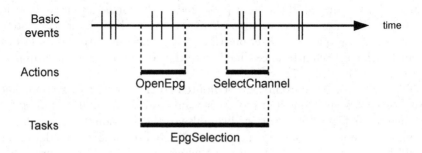

**Fig. 3.2.** Sequences of basic events describe user actions, which again can be combined into tasks.

"open the program guide", executes the same action, but produces a different sequence of events. Therefore, an interpretation of basic events is required to identify meaningful subsequences. Once user actions have been identified, a higher-level description of user behavior becomes possible. As can be seen in Figure 3.2, we describe user behavior as a hierarchy of observations. Basic events form user actions, which in turn constitute the building blocks for tasks. In the remainder of this chapter, we present approaches for recognizing user actions from basic events and describing higher-level user behavior by means of tasks.

### 3.1.2 Behavior as Actions and Data

User behavior in interactive systems consists of actions and data. Different subsequences represent user actions. On the one hand, the user may choose from a number of possible input methods. They represent modalities that create different sequences, such as mouse, remote control, speech input, and so on. On the other hand, different methods trigger the same action within a modality, such as different buttons on the remote control. Thus, a set of different event sequences describes a user action. The functionality provided by an interactive system defines the list of possible user actions. For instance, a TV system allows the user to change the channel and the volume. The user performs action sequences during the interaction. These sequences form tasks that describe a higher-level user-system interaction. For instance, the task of route destination entry in personal navigation devices comprises the steps of entering a city, a street, and a house number and is defined by a number of basic actions. Moreover, the order of these actions is not fixed, but the user may decide about the order of actions. User modeling algorithms detect user actions, describe higher-level user behavior, and predict user actions or sequences of actions.

In addition to the meaning of an action, data defines this action further. For example, if a user selects the "BBC" channel, the "SelectChannel" action has the name of the channel "BBC" as a parameter. This data enables the interactive system to describe user preferences. For this purpose, it extracts the data and employs user modeling algorithms on it. The knowledge about user preferences is built by the user modeling component by modeling the data associated with actions. For instance, if the channel name "BBC" is the most frequent parameter of the "SelectChannel" action, the user modeling component infers that "BBC" is the user's favorite channel. The research field of data mining has produced an extensive number of algorithms that are applicable to user modeling (see Chapter 2). An important requirement for user modeling algorithms is transparency, thus allowing the user to comprehend and anticipate the results of the computation. Therefore, straightforward algorithms may be a sufficiently good choice. One example is the "most frequently used" (MFU) algorithm, which picks the element from a list that was selected most often by the user. In addition, domain-specific algorithms may be better suited for producing good predictions than generic algorithms, since they may consider domain knowledge. For instance, different algorithms specifically for modeling the user's TV preferences have been presented (see Ardissono et al., 2003 and Vildjiounaite et al., 2008).

Thus, user modeling comprises both actions and data. For this purpose, different kinds of information are derived from a specific event sequence. For example, user modeling algorithms predict user actions based on previous actions and the user's favorite channels based on previously selected channel names. Algorithms for modeling user preferences from observed data often are domain- and application-specific. However, user actions may be described in a domain-independent way. We present generic algorithms for describing actions in the remainder of this chapter.

## 3.2 Recognizing User Actions in Event Sequences

In order to facilitate a comprehensive description of user behavior, meaningful actions first have to be extracted from a sequence of basic events. Each user action, such as selecting a channel in a digital TV system, is represented by different sequences of events. In the following, we present an approach we devised for extracting these subsequences from the user-system interaction and detecting user actions.

Instead of using machine learning, the developer may extend the interactive system as to emit specific events to indicate which action was performed. For instance, the developer may connect the events to interface elements, such as graphical buttons. In this case, an observation of the action event indicates that the user executed the respective action. However, this approach requires extensive work by the developer. In addition, the developer may align actions with a mapping between system events and user actions. Whenever a specific system event is observed, the user triggered the action associated with that event. However, the developer has to ensure that the mapping includes all relevant events and actions. Both approaches rely on a definition of user actions by the developer. In addition to the extra work, the system designer may easily miss actions and therefore produce an incomplete definition.

Data driven approaches recognize user actions based on sample data. These approaches collect training data covering all actions. Sample data may be collected by annotating sessions that were recorded with test users. After the annotation has been performed, machine learning techniques extract user actions from the interaction. Since the sequences that describe user actions are quite different in length, hidden Markov models and artificial neural networks are not well suited for this problem. Instead, we introduce an approach for describing user actions by means of probabilistic automata in the following. After an introduction of probabilistic automata, we discuss the application of this tool to the problem of recognizing user actions.

### 3.2.1 Probabilistic Deterministic Finite-state Automata

Finite state automata consist of a number of states $Q$ and an alphabet $\Sigma$. At the beginning, one state that is marked as initial is activated. A number of final states terminate the automaton. Transitions between states are defined as $\delta : Q \times \Sigma \times Q$. That is, a transition $\delta$ connects two states with a symbol of the alphabet $\Sigma$. For example, the transition $\delta_a : Q_x \times \sigma_w \times Q_y$ defines that, with $Q_x$ being the active state in an automaton and the symbol $\sigma_w$ being observed, the automaton activates state $Q_y$. A deterministic automaton allows only a single transition between two states with a specific symbol (Equation 3.1).

$$\forall q \in Q, \forall a \in \Sigma, |\{q' : (q, a, q') \in \delta\}| \leq 1 \qquad (3.1)$$

Probabilistic deterministic finite-state automata (PDFA), described in detail by Vidal et al. (2005a,b), extend automata with probabilities. For this purpose, each transition in a PDFA possesses a transition probability $P(\delta)$ and every state has a final-state probability $F(q)$. The sum of all transitions with a common source state and the final probability of the state equal to 1 (Equation 3.2).

$$\forall q \in Q, F(q) + \sum_{a \in \Sigma, q' \in Q} P(q, a, q') = 1 \qquad (3.2)$$

In addition, exactly one initial state is active at the beginning. A graphical representation of an example PDFA is given in Figure 3.3: In the initial state $q_0$, the symbol $x$ triggers a transition to state $q_1$ and occurs with a probability of 0.35, whereas the transition triggered by the symbol $y$ to state $q_2$ occurs with a probability of 0.65. State $q_1$ has a final-state probability of 0.1, i.e., the automaton terminates in state $q_1$ with a probability of 0.1.

A state acceptor represents an application of the state automaton to determine if an automaton matches a given sequence and, if so, it computes the probability. The acceptor starts at the initial state and processes the sequence of symbols. For each symbol, the acceptor searches a transition in the active state that matches the current symbol. If a transition is found, the acceptor advances to the target state of the transition. If no matching transition is found, the automaton does not accept the sequence and terminates. Otherwise, this step is repeated for the remainder of the sequence. The acceptor matches the sequence if the automaton is in a state with a final-state probability greater than 0 at the end of the sequence. The probability of the accepted sequence for an automaton is computed by multiplying the transition probabilities of all transitions taken. Therefore, the probability for the sequence $(x, z)$ in the automaton in Figure 3.3 is 17.5 %.

Learning PDFAs from a list of sequences works as follows (see Vidal et al., 2005b). An automaton initially consists of only the initial state. The following training procedure is performed for each sequence in the training data. Starting at the initial state, the training algorithm seeks a matching transition for the current symbol of the sequence in the active state. If no matching transition is found, a new state and a transition to the new state with a weight of 1 are added to the automaton. Otherwise, the weight of the existing transition is increased. After all sequences have been processed, the probabilities are computed from the weights by dividing the weight of a transition by the sum of the weights of all transitions that have the same source state.

Thus, PDFAs provide an approach based on finite-state automata for sequence classification. The approach relies on training sequences to learn a set

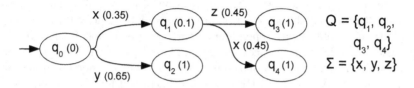

**Fig. 3.3.** An example of a probabilistic deterministic finite-state automaton.

of automata. In the following section, we discuss the application of PDFAs to the problem of recognizing user actions in interactive systems.

### 3.2.2 Describing User Behavior using PDFAs

In this section, we present probabilistic automata as a solution to the problem of recognizing user actions. For this purpose, the basic events from the user-system interaction constitute the alphabet $\Sigma$. At design time of the interactive system, we learn a set of PDFA matchers from annotated training data that has been collected in test sessions. The interactive system then applies the matchers to live interaction events to recognize user actions at runtime. An overview of the workflow is given in Figure 3.4. First, log data is collected, for instance by means of user tests. This data is then annotated by assigning action names to sections of the recordings. Next, the interaction sequences that correspond to the different actions are extracted. Probabilistic matchers are trained based on the extracted sequences. In the following, we discuss the individual steps in detail.

The first step of the learning process depicted in Figure 3.4 collects and annotates training data. This data has to cover all user actions that should be recognized. Moreover, the log in the training data has to be identical to the log that is produced during the interaction of a user with the interactive system. For instance, training data is collected in test sessions with users. If no test users are available, the system designer performs sessions with the interactive system. The system records the events and writes them to log files. Once the training data has been collected, an annotation by means of labeling is performed. For this purpose, sections in the interaction timeline are identified in which a user performs actions. These sections are labeled with the name of the action. In order to support the annotation process, we extended a graphical evaluation tool with annotation facilities (Wesseling et al., 2008). The tool displays a user session in a graphical timeline view. Sections that correspond to a user action are selected visually. The annotation collects different representations of each user action, for example because the action was triggered using different modalities. The list of user actions is either defined before the annotation or constructed implicitly by adding actions as needed. Consistency and precision of the annotation are prerequisites for a well performing matcher. After the annotation of the training data, all sequences that represent the same action are extracted. Specific sequences occur multiple times or only once, depending on the user's behavior. Each action is represented by different sequences, because different modalities emit different sequences for the same action or the user has different methods available to trigger an action. A separate PDFA acceptor is trained for each action from the extracted event sequences using the algorithm presented in Section 3.2. The PDFA learning concludes the training process, which is performed at design time of the interactive system.

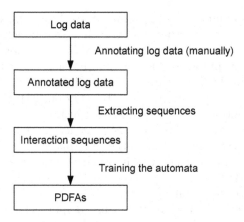

**Fig. 3.4.** The workflow of the probabilistic automaton approach for recognizing user actions.

Adaptive interactive systems employ the matchers at runtime to recognize user actions. For this purpose, the system loads all PDFA acceptors at startup and submits events to the matchers. Each matcher recognizes one specific user action. When entering a state with a final-state probability greater than zero, a matcher recognizes a user action. All actions that occur in a sequence are collected and the most likely action is selected if more than one action was recognized. However, the length of an accepted event sequence is not known in advance. Therefore, the matcher uses a timeout to collect different sequences and selects the one with the highest probability after that timeout. In addition, a threshold suppresses wrong recognitions with low probabilities, which may occur due to errors in the annotations. Users may perform actions using different modalities, such as remote control and speech input in the example. In doing so, different event sequences represent a single action. All sequences that represent an action are extracted and used to train a set of PDFA matchers. Figure 3.1 presents an example sequence of basic events from a digital TV system. In lines 1–4, the user opens the electronic program guide. Therefore, the system designer marks the respective sequence with the name of the action, which is "OpenEpg". Lines 5–7 represent the "SelectChannel" action.

Preprocessing of the log data is necessary for several reasons. The individual preprocessing steps depend on the requirements of the user modeling algorithm and the adaptations. A number of useful steps are discussed in the following. First, some event types are not required for differentiating user actions and are therefore filtered out to prevent a negative impact on the recognition. Different actions have the same appearance in the log data and the meaning depends on the context. For instance, a remote control button may have different meanings in different screens, e.g. opening the result list in one screen or recording an entry in another one. The actions become distinguishable by adding context information to the events, such as the name of the active view or state. Other events, such as events defining global commands like scrolling in a list or changing the volume, do not have context information. Moreover, preprocessing may add missing information that is not present explicitly in the event parameters. For instance, the modality can be inferred automatically from event types. In addition, PDFA matchers recognize user actions, but do not address data. For instance, a matcher recognizes the "SelectChannel" action, but does not extract the name of the selected channel. Manual extraction rules and other information extraction techniques (see Appelt, 1999) add the respective information to events. For example, an extraction rule adds the name of the channel to a speech recognition event based on the recognition result.

We present an evaluation of the PDFA-based approach for recognizing user actions in Chapter 6, which proves the feasibility of the approach. Therefore, PDFAs are a viable means for describing user actions in interactive systems. They present a more intuitive and less error-prone approach than a manual description of user actions. Moreover, creating the matchers does not require a detailed knowledge of the inner workings of the interactive system. In addition, the matcher may be used in an evaluation to extract user actions in a set of log files automatically. For this purpose, a part of the log sessions are annotated and the resulting matchers are applied to the remaining log files. Thus, these matchers may support the investigation of user behavior in an evaluation.

## 3.3 Modeling Tasks in Interactive Systems

Once user actions have been identified in a sequence of events, a higher-level description of user activities becomes feasible. The higher-level description defines which actions a user can perform in a specific interaction state as well as their order. Based on this description, the interactive system for instance recommends actions or detects user problems. In the example from the introduction of this chapter, the interactive system has recognized the "OpenEpg" and "SelectChannel" actions by means of the PDFA approach presented in the previous section. Based on an explicit representation of possible user actions, the interactive system knows that the user may perform the "SelectGenre" or "OpenResults" actions next. Such an explicit representation of user actions is called task model.

Tasks are hierarchical descriptions of activities that aim to accomplish a goal. Task models present a formalism for defining which tasks a user can perform in an interactive system. The process of creating a set of tasks for an interactive system is called task modeling. Task modeling has originally been intended for the development process, for instance for automatically generating interfaces, and for the evaluation of interactive systems (Paternò, 2001). In the following, we use task models as an approach for describing higher-level user activity at runtime of an interactive system. Task models enable an adaptive interactive system to observe the user-system interaction and derive adaptation information accordingly. For instance, the adaptation may recommend a specific action to the user.

In the remainder of this section, we present a novel approach that uses task models at runtime of adaptive interactive systems to track the user-system interaction. We introduce the concept of task models and present an approach for applying them at runtime. Next, we discuss how to derive different kinds of information from the task model, such as a prediction of the next user action or the proficiency of the user with the interactive system. We present how this information serves as a basis for deciding about possible adaptations.

### 3.3.1 Task Modeling

In the following, we introduce task models are a method for describing user behavior. A task defines the activity required to accomplish a specific goal. Activity is defined by means of a hierarchical arrangement of user actions. Different relations exist between these actions. Two actions are executed sequentially (action A enables action B) or alternatively (user executes action A or action B). For instance, task X may be defined such that the user has to perform actions A, B, and C in sequence to achieve goal Z. A task model defines all actions a user may execute while working on a task. However, a subset of these actions suffices to finish the task. For instance, the user may select among alternative actions or omit optional actions. A plan defines a strategy a user follows to accomplish a goal and a plan is a subset of a task.

In order to describe a task, the individual actions that can be performed during that task are identified first. Thereafter, the task model is constructed using actions as building blocks. The process of identifying the individual actions and arranging them accordingly is called task analysis. Different techniques exist. Hierarchical Task Analysis (HTA; Annett and Duncan, 1967) originates from the domain of industrial training. As a top-down analysis, HTA starts with a goal and breaks the goal down recursively into tasks and subtasks that have to be performed to accomplish the goal. HTA focuses on physical and observable actions. In HTA, plans represent paths through the hierarchical structure, because the user does not have to perform all tasks for each plan. HTA includes a notation for tasks. GOMS (Goals, Operators, Methods, and Selection Rules; Card et al., 1983) presents another approach for task modeling. Further methods have been derived from GOMS (see John, 1995), such as Keystroke Level Modeling (KLM) and Cognitive, Perceptual, and Motor GOMS (CPM-GOMS). GOMS describes user activity as goals, operators, methods, and selection rules. A goal defines the purpose of the activity and a number of operators are available to achieve that goal, such as menu selection or mouse movement. In addition, a number of methods specify action sequences similar to plans. Methods are defined hierarchically down to the level of keystrokes. Finally, a number of selection rules select the most appropriate methods. Kieras (2003) presents a top-down approach for building

GOMS models that starts with a goal and decomposes the task continually, until arriving at primitive operators at the bottom. Due to the complexity of the models, different tools have been created for constructing GOMS models (see Baumeister et al., 2000). However, there is no standard notation. GOMS is used mainly during the evaluation of interactive systems and allows system designers to evaluate an interactive system with a GOMS model instead of real users. For instance, the time required for a task is computed by assigning interaction times to the individual actions and running a simulation with the GOMS model.

Different notations for task models exist, with one of the most well-known formalisms being ConcurTaskTrees (CTT; Paternò et al., 1997). A CTT model consists of a hierarchy of basic tasks, which are user tasks, application tasks, interaction tasks, or abstract tasks. The user performs a "user task" without interacting with the system, whereas the system performs an application task without user interaction, e.g. a database query. Interaction tasks describe a communication between the user and the system, for instance a mouse click or a speech interaction that trigger system functions. Abstract tasks are composed of a combination of the other task types. Different temporal operators define the order in which the tasks are executed. The most important operators are *interleaving* (an arbitrary order), *enabling* (one action enables another), *iteration* (repetitive execution), and *optional task* (optional execution).

The task model we use in this work is similar to CTT models, but we apply them at runtime to derive adaptation information. We limit the types to system and interaction tasks, because user actions without system interaction cannot be observed, as discussed in Section 3.1. User actions constitute the building blocks of the task model. They are grouped using sequences (i.e., the actions occur in the specified order) or alternatives (i.e., only one of the listed actions is performed). Action groups may be optional or iterative, i.e., they possibly occur more than once. In addition, actions may be marked as final if they terminate the current task. Final actions may reference other tasks to enable the task model to determine directly which task becomes active after the current task.

We use a set of task models to describe the different tasks in an interactive system. These correspond to different goals and parts of the system. Thus, the task models for a digital TV system include watching TV, browsing the teletext, or using an EPG. An example of a task in a graphical, statechart-like notation is given in Figure 3.5. The notation includes user actions, final user actions, and decision states, in which the user decides which action to perform next. After selecting a set of criteria, such as time or channel, in the EPG of a digital TV system, the user opens the result screen and thus finishes the task by means of the "Show results" action. If the user decides to open a filter selection menu, for instance by means of the "Go to time selection" action, he or she either selects a value (e.g. using the "Select time" action) or cancels the selection (using the "Cancel" action). Thus, a task model provides a comprehensive description of higher-level user activity.

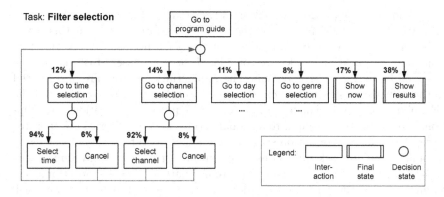

**Fig. 3.5.** An example task model that describes the user's activity when selecting a set of filter criteria in an EPG.

### 3.3.2 Constructing Task Models

In the following, we discuss the construction of task models in more detail. Approaches for an automatic definition of task models exist, for instance by deriving the model from an existing application model or log data. However, a manual revision is required due to the high amount of domain-dependent information that only a domain expert knows.

Garland et al. (2001) present a semi-automatic approach for constructing task models. They regard building task models as a complex endeavor that has to be performed at least in part by domain experts. Rather than constructing a task model automatically, the tooling supports a domain expert in building the task model. The process starts with a set of example tasks that are used by a machine learning engine to infer task models from annotated examples. The domain expert continuously refines the examples and the task model, until the model corresponds to the interactions of a user with the system. The individual steps require manual work by a domain expert.

Our task modeling approach also includes automatic support for building initial task models, which are then refined by a domain expert. For this purpose, we present two techniques. First, an initial task structure may be generated from the system model, if a mapping between the system description and user actions is available (e.g. by mapping events to user actions). Second, a coarse task structure may automatically be extracted from action sequences in recorded user sessions, if annotated recordings are available. However, both approaches only produce coarse task models that serve as a starting point for a manual revision.

The task model notation used in this work uses an XML format. Although these files may be created with a text editor, tool support significantly eases this complex task. The ConcurTaskTree Environment (CTTE, Mori et al., 2002) is a graphical editing tool for the CTT notation. In order to create task models for the adaptation framework presented in this work, we provide a converter from CTTE models to the task model format used in this work. This converter only supports a subset of the CTT notation. Only "interaction" and "application" tasks are considered, because "user" tasks cannot be observed by an interactive system. The sequential operators are also limited to the "Choice" operator, which is mapped to an alternative action, and the "Enabling" operator, which is mapped to a sequence. In addition, the "Optional" operator is translated to optional actions and the "Iteration" operator to repetitive actions or action sequences. Further operators are not necessary to build complete task models used at runtime of interactive systems. Constructs of the task model format that are not part of the CTT notation, such as final actions, are added by the domain expert after the export. The task model is loaded at runtime by an interactive system to track the user-system interaction. In the following section, we discuss how an interactive system uses task models at runtime and how adaptations are triggered by a task model-based description of the interaction.

### 3.3.3 Task Models in Adaptive Systems

Task models have mainly been used as a support for the development and the evaluation of interactive systems. In the following, we introduce a novel approach we developed to exploit task model information at runtime of interactive systems to serve as an adaptation trigger. At the beginning of a user session, the system loads the task models. The individual building blocks of the task models correspond to user actions. Whenever the system observes a user action, for instance by means of the PDFA approach discussed in Section 3.2, the task model status is updated accordingly. For this purpose, the task modeling component keeps a list of active tasks and stores the active state of each task. If an observed action is the initial one of an inactive task, the respective task is activated. For this purpose, it is added to the list of active tasks and the initial state of the model is activated. If an action is a possible next step of an active task, the status of the active task is updated. If an activated state is a final state, the task is closed. If a step is not allowed according to an active task, the respective task is terminated. In addition, conditions increase the validity of the task model, such as the current graphical screen or speech state.

An example task model that describes the digital TV system presented in the introduction is given in Figure 3.5. This model illustrates the task of selecting filter criteria in an EPG and opening the result screen. When the interactive system starts, the task is not active. Once the user opens the EPG by means of the "Go to program guide" action, the "Filter selection" task is activated. When the user opens the time menu using the "Go to time selection" action, only the "Select time" and "Cancel" actions are available. The task is finished when the user opens the result screen by means of the

"Show results" action. The "Show results" action may be the start action of another task.

By tracking user behavior with a task model, the interactive system can anticipate user actions based on the list of valid actions in the current state. However, probabilistic information improves the utility of a task model considerably and facilitates different applications of a task model. Different methods exist for adding statistical information to a task model. First, the system designer annotates the task model with the expected probabilistic information. For instance, the system designer could expect that a user selects action A with a 60 % likelihood and action B with a 40 % likelihood. This information does most likely not reflect actual probabilities, but serves as a starting point. Second, the task model is trained from either recorded or live log data. For this purpose, the task model tracks the interaction, as discussed previously, and increases a counter in the model for every action it observes. The probabilities are derived from the respective counters. The task model is trained either from training data of multiple users, of a subgroup, or of a single user, depending on the intended application of the information. The example task model in Figure 3.5 includes statistical information. After having entered the EPG screen, users select the "Go to time selection" action with a probability of 12 % and the "Show results" action with a likelihood of 38 %.

In the remainder of this section, we discuss different applications of task models we devised for adaptive interactive systems. The aim is to derive information to trigger adaptations, such as user problems or predictions. For example, a prediction of the next user action triggers an adaptation that highlights specific interface elements.

**Predicting the Most Likely Next User Action**

One application of a task model is to predict the most likely next user action. An adaptation component uses this information to provide assistance accordingly, for instance by means of adaptive help messages or by emphasizing interface elements. The prediction relies on statistical information in the task model. For this purpose, this model tracks the interaction and therefore is aware of the current interaction state. The user can perform only a limited number of actions in each state. The task model predicts the most likely next user action by selecting the one with the highest probability value.

For example, in the task model of Figure 3.5, the user enters the EPG of a digital TV system by means of the "Go to program guide" action. According to the task model, the "Go to time selection" action is performed with a 12 % likelihood, the "Go to channel selection" action with a 14 % likelihood, and the "Show results" action with a 38 % likelihood, etc. Therefore, the task model predicts the "Show results" action, since this action has the highest probability.

Which action the task model predicts depends on the training data. If the model is trained with data from a group of users, the prediction reflects the preferences of the group. For example, a task model recommends the most common action to novice users by training the model with data from a group of different users. On the other hand, a task model reflects the interaction style and preferences of an individual user if the training data is taken only from the current user.

## Recommending Unknown Actions

In addition to recommending the most likely action, the task model can recommend actions that the user has not performed yet. For this purpose, the model records the interaction of the user and thus knows the actions a user performed. A comparison of the interaction of the current user with the interactions of other users improves the recommendation. For this purpose, the system employs two task models, one with the statistical information of the current user and one with statistical information of other users. Both models track the user-system interaction in parallel. The interactive system compares the probability of the predicted actions of both models. If the probability of the model that represents a user group is higher than the one of the model that describes the user, the action predicted for the group may be recommended to the user.

An example shall illustrate this approach. The task model in Figure 3.5 contains an action called "Show now", which opens the list of TV shows that are currently on the air. The group of all users executes the "Show now" action with a 17 % probability. If the current user however had not used the "Show now" action at all or with a probability significantly lower than 17 %, the task model could recommend this action, assuming that the user does not know about the respective feature. Since the probability of the "Show results" action (21 %) is higher, the model would never recommend the "Show now" action by only considering the most likely action.

## Finishing the Current Task

The task model may present a sequence of actions to the user for finishing the current task. For this purpose, the model computes a plan for ending the current task by regarding the model as a graph and finding the shortest path between the active state and a final state. For example, Dijkstra's shortest path algorithm (Dijkstra, 1959) is suitable for computing the shortest sequence of actions to finish the current task. Instead of using the weight of the individual edges for finding the shortest path, the actions with the highest probabilities are selected.

The adaptive interface uses the recommended sequence of actions to show help that explains the sequence needed to finish the current task or to execute the action sequence automatically. The task model in Figure 3.5 illustrates this approach. If a user performed the "Go to time selection" action, the shortest path algorithm generates the sequence ("Select time", "Show results") to finish the "Filter selection" task. The algorithm is more useful in combination with more complex task models and long action sequences.

### Detecting User Errors and Problems

A task model also reveals user problems in the interaction. For instance, the user might hesitate in a task or frequently switch between tasks without finishing a task. First, time is an indicator of user problems. If a user hesitates while interacting with the system, he or she might be in need of help. However, there are times of inactivity in the interaction, for instance when the user is watching TV or working on tasks outside of the interactive system. The task model recognizes hesitation by comparing the interaction times of the current user with interaction times of other users. For this purpose, the task model investigates the waiting times at the decision states. In addition to recognizing user problems, the task model can infer the user's proficiency, for instance by connecting short waiting times to expert users.

Another indication of user problems is frequent task switching. If the user does not finish tasks, but switches to other tasks continuously, the user may need support, for instance by means of help messages. The user finishes a task by executing an action that is marked as final action. Therefore, the task model infers user problems if it observes that a user frequently starts new tasks without performing one of the final actions of the previous task. In addition, if a user needs more interaction steps to finish a task than the average user, the interactive system assumes that the user needs help. For instance, of a user switches frequently between the "filter selection" (shown in Figure 3.5) and "watch TV" (not shown) tasks, a help message that explains how to use the filter menu helps the user.

Thus, we introduced a novel approach for describing higher-level user behavior in adaptive interactive systems. We presented task models as a viable means for this description. These models track user behavior, compute predictions of user actions or action sequences, and derive further information about the user-system interaction, such as detecting problems or errors.

## 3.4 Predicting User Actions

An adaptive interactive system anticipates user behavior, for instance by predicting a user action. Such an interface may highlight interface elements related to predicted user actions or offer shortcuts for a sequence of user actions. In this section, we introduce algorithms for predicting either a single user action or a sequence of user actions. We adapted these algorithms for the use in adaptive interactive systems from other domains.

### 3.4.1 Predicting One User Action

A prediction of the next user action enables an interactive system to trigger adaptations. For instance, the system provides help for the action or highlights interface elements that are connected to the anticipated action. We introduce two approaches for predicting a single user action in this section, an algorithm based on Markov chains and an algorithm that exploits a task model. An evaluation of both algorithms is presented in Chapter 6.3.

#### Task Model-based User Action Prediction

An algorithm for predicting a user action by means of probabilistic information in a task model is presented in Section 3.3. A task model with probabilistic information tracks the user-system interaction and produces a list of actions the user can execute. The model selects the action with the highest probability as a prediction. An example of an action prediction based on task model information is presented in Section 3.3.

However, this approach always recommends the same action at a specific state, although the actual actions of a user depend on the interaction history. For instance, the task model in Figure 3.5 always recommends the "Show results" action, because this action has the highest probability of 38 %. Instead of selecting the results screen directly, a user selects filter criteria, such as channel or time, before opening the results screen. Therefore, the prediction is improved by recommending the "Go to day selection" or "Go to channel selection" action first. When the user enters the selection screen again, the interactive system should recommend the "Show results" action. Thus, including the interaction history into the prediction improves the usefulness of the recommended action. In the following, we present an action prediction algorithm that incorporates past user actions. For this purpose, we adopted an algorithm for link prediction from the domain of adaptive hypertext to the task of user action prediction in interactive systems.

#### Using Markov Chains for User Action Prediction

A prediction of a user action becomes more useful by incorporating the interaction history. In this section, we present an algorithm based on Markov chains that considers the interaction history. Markov chains are a statistical tool for modeling sequences and have been applied to adaptive interactive systems. For example, Zhu et al. (2002) and Sarukkai (2000) use Markov chains to model user behavior for predicting links in adaptive websites. We adapted the Markov chain-based approach to the domain of action prediction in interactive systems by including domain knowledge.

Markov chains exploit the Markovian assumption, which states that a symbol in a sequence only depends on the previous symbol. A sequence S consists of a number of symbols $x_n$ and a transition probability matrix A stores the probabilities $a_{ij}$ of symbol $x_i$ being followed by symbol $x_j$. A prediction of the next symbol is computed by comparing all probabilities in the column of the matrix that is associated with the current symbol and selecting the one with the highest probability. A Markov chain that only considers sequence of length one is called $1^{st}$ order Markov chain. Markov chains that consider longer sequences are called $2^{nd}$ order or higher order Markov chains. For instance, with an alphabet of ("A", "B"), a $1^{st}$ order Markov chain stores probabilities for "A" and "B", whereas a $2^{nd}$ order Markov chain stores probabilities for the sequences "AA", "AB", "BA", and "BB". With an alphabet of 40 symbols, there already exist 1600 combinations with a $2^{nd}$ order Markov chain. Since the computational complexity of higher-order Markov chains increases significantly, higher-order Markov chains are not well suited for devices with limited computational resources, which often applies to interactive systems.

Therefore, we only consider first order Markov chains in this work. The interaction history is included in a different way. Sarukkai (2000) presents an approach for link prediction that stores the user's interaction history and includes the history into the prediction. In the following, we present how we adapted this algorithm to the domain of user actions. The user actions constitute the alphabet and an index is assigned to each action. For example, the actions $a$, $b$, and $c$ receive the indices 1, 2, and 3. Actions are represented as vectors and define a probability distribution of the action. For instance, action $a$ with index 1 is represented by means of the vector $s_a = [1, 0, 0]$. The probabilities of other actions are 0 in this case. The fact that the actions $a$, $b$, and $c$ occur with the same probability of $\frac{1}{3}$ is expressed by the vector $s = [\frac{1}{3}, \frac{1}{3}, \frac{1}{3}]$. A matrix $A$ represents the transition probabilities between the actions. The probabilities of the next actions $s(t)$ are computed by multiplying the matrix with the vector that represents the current observation $s(t-1)$ (Equation 3.3). This equation exploits the Markovian assumption.

$$s(t) = s(t-1) * A \qquad (3.3)$$

The interaction history is not yet included, but the next action is predicted based on the previous one. The history $h$ stores the last N action vectors. If an action is observed, the new action is added to and the oldest action removed from this vector. In this work, a number of N=5 observations is used. For example, the action sequence [a, b, c, a, a] is represented by means of the vector [[1,0,0], [0,1,0], [0,0,1], [1,0,0], [1,0,0]]. A prediction for the last N actions is computed and the individual results are added up, with older actions receiving a lower weight. The matrix $A^n$ produces a prediction for the $n^{th}$ action in the sequence. The weight of older actions is decreased by means of a dampening factor $d = (1.00, 0.50, 0.30, 0.25, 0.20)$. The result vector is normalized in order to produce probabilities. Equation 3.4 presents the prediction algorithm that incorporates the interaction history.

| | Observed<br>user action | Task model<br>prediction | Markov chain<br>prediction |
|---|---|---|---|
| | \<open EPG\> | \<open EPG\> | \<open EPG\> |
| 1 | Go to day selection | Go to day selection | Show results |
| 2 | Select day | Select day | Select day |
| 3 | Go to time selection | Go to time selection | Show results |
| 4 | Select time | Select time | Select time |
| 5 | Go to channel selection | Show results | Show results |
| 6 | Select channel | Select channel | Select channel |
| 7 | Show results | Show results | Show results |

**Table 3.1.** Observed user actions and predictions by the task model algorithm and the Markov chain algorithm.

$$s(t) = \sum_{j=1}^{N} d(j)h(t-j)A^j \tag{3.4}$$

The transition matrix is trained using either events from the current interaction to reflect the characteristics of an individual user or recorded sessions of a group of users to represent the characteristics of a user group. First, a transition count matrix $T$ is created from a sequence of observations. If action $x_j$ is followed by action $x_i$, the field $t_{ij}$ of the matrix is incremented. The probability transition matrix $A$ is derived from matrix $T$ by dividing each cell by the sum of the respective row.

However, the Markov chain prediction sometimes produces predictions that are not valid according to the task model due to the statistical nature of the prediction. Therefore, we use a combination of the Markov prediction and the task model information to ensure the validity of the prediction. For this purpose, the task model tracks the interaction and the Markov chain algorithm produces a prediction vector. Thereafter, the predictions that are not valid according to the task model are removed.

Table 3.1 gives an example of the predictions of the task model algorithm and the Markov chain algorithm, based on the task model given in Figure 3.5. As can be seen, the task model algorithm always predicts the "Show results" action in the EPG main screen, whereas the Markov chain algorithm predicts the "Go to day selection" action correctly in line (3). However, the Markov algorithm predicts the "Show results" action in line (5), whereas the user selects another filter criterion first. An evaluation of the two algorithms is presented in Chapter 6.3. Thus, the Markov chain algorithm outperforms the task model algorithm due to the combination of statistical prediction and domain knowledge.

### 3.4.2 Predicting a Sequence of User Actions

In addition to the prediction of a single action, predicting a sequence of actions allows an adaptive interactive system to offer shortcuts for executing the predicted sequence more quickly or conveniently. For example, if the user frequently selects a specific set of filter criteria, the system may provide shortcuts accordingly. In this section, we present an approach for predicting sequences of user actions that exploits sequence mining. Interaction pattern mining extracts a set of frequent action sequences from an interaction and transforms them into rules. These rules enable a prediction of action sequences.

Instead of using supervised learning techniques, such as case-based reasoning, neural networks, or probabilistic automata, sequence mining techniques extract frequent episodes from a longer sequence without the need for training data. Agrawal and Srikant (1995) present different algorithms for sequence mining and illustrate these algorithms with an example from the domain of customer transactions. The algorithms exploit the a priori property. It states that an episode with $k$ symbols can only be frequent if all sub-episodes are frequent. Another approach presented by Mannila et al. (1997) also exploits the a priori property. Liu et al. (2003) employ this algorithm in adaptive office software that learns frequent formatting options and presents them as shortcuts to the user. Similarly, Heierman and Cook (2003) use sequence mining in the domain of intelligent home automation. In this work, we use an algorithm that is adapted from Mannila et al. to extract frequent action sequences from the user-system interaction. This information may for instance be used to present them as shortcuts to the user. An example system and an evaluation of the interactive system are presented in Section 6.4.6.

### An Algorithm for Mining User Actions

We introduce a sequence mining algorithm, which is based on an algorithm presented by Mannila et al. (1997). We start with a definition of a number of terms. An episode is a subsection of an event sequence and a frequent episode appears more often than a previously defined threshold. An episode X is a sub-episode of sequence Z if X represents the beginning of Z. In this case, Z is a super-episode of X. A frequent episode is closed if there is no longer episode in a set that contains it as a sub-episode. For instance, in the set of episodes ($AB$, $ABC$, $ABD$), $AB$ is a sub-episode of $ABC$ and $ABD$. Thus, the episode $AB$ is not closed, because longer episodes exist that have the sequence $AB$ as a prefix. A rule consists of two episodes, which are called trigger and implication. The trigger is a sub-episode of the implication. If a trigger appears in a sequence, the remaining part of the implication follows with a given probability. For instance, the rule $AB/CD$ has the episode $AB$ as a trigger. When the $AB$ sequence is observed, the rule predicts the sequence $CD$.

The algorithm regards user actions as symbols. For this purpose, it processes a sequence of user actions, for instance a live user-system interaction or a log file. This algorithm comprises the following steps: mining frequent episodes, selecting closed episodes, and creating rules. First, a set of frequent episodes is extracted from a sequence. For this purpose, the algorithm finds episodes of length k (k ≤ 2) by combining episodes of length k-1. Thereafter, it filters results by means of the a priori property and derives probabilistic information. Pseudo code of the algorithm for mining frequent episodes is given in Figure 3.6. The algorithm moves a sliding window with an increasing length over the sequence. If the beginning of the episode in the window is a frequent episode (or is not longer than a minimum length), this episode is added to a list of candidates. The algorithm counts the occurrences of the individual episodes. At the end of one iteration, all episodes that occur with a frequency higher than a threshold *minOccurrence* are added to the list of frequent episodes. In this work, a value of two is used as the *minOccurrence* threshold. The algorithm terminates either at a given upper length *lookahead* (values of five to 20 are used for this work) or if no more episodes are found in the current iteration. The algorithm is based on the INTEMTM algorithm (Mooney, 2007), which again exploits the a priori property and the windowing method used in the WINEPI algorithm presented by Mannila et al. (1997).

```
function findFrequentEpisodes (sequence S, lookahead, minOccurence):
    frequentEpisodes = []; length = 1; found = false

    while length < lookahead and found:
        candFreqEp = []; found = false

        for (i=0; i<S.length - length; i++):
            create new episode ''a'' with length ''length'',
              starting at index ''i''
            if beginning of ''a'' is frequent episode:
                increment counter of a

        for (episode a in candFreqEp):
            if counter of a > minOccurence:
                add a to episodes of length ''length''
                found = true
        length = length + 1
    return frequentEpisodes
```

**Fig. 3.6.** Pseudo code of the algorithm for mining frequent interaction episodes.

After a set of frequent episodes has been extracted from the interaction sequence, all non-closed ones are removed. For instance, the episode *AB* is removed from the set (AB,ABC) since it is a sub-episode of episode *ABC*. Finally, rules are generated from the set of frequent episodes. The confidence of a rule is the probability that a trigger turns into the implication. The confidence is computed using Equation 3.5, with $\alpha$ being a sub-episode of $\gamma$ in a sequence $s$ and $fr$ the frequency of the episodes.

$$conf(r) = fr(\alpha, s)/fr(\gamma, s) \qquad (3.5)$$

Therefore, if an episode $\alpha$ turns into episode $\gamma$ four out of five times, the confidence is $\frac{4}{5}$. The confidence of the sub-episodes is computed for every closed episode, starting at a length of one. The rule is generated once the confidence reaches a threshold. Thus, the algorithm selects the rule with the shortest trigger for a fixed confidence level. Values of 0.5 to 0.7 are used as confidence threshold. However, the confidence level may be connected to the length of the sequences to better support sequences of different lengths.

Rules predict event sequences by observing an event sequence and matching the observations with the rule triggers. If a trigger is observed, the remainder of the rule implication follows with a given probability. The length of the prediction depends on the length of the rule and the length of the trigger. In order to produce a prediction of an action sequence instead of a single action, the difference in length between the trigger and the rule length has to be at least two. If a minimum length of the prediction is required, rules with shorter predictions are omitted. Moreover, episodes can contain repeating parts. For example, the sequence $ABCBCD$ contains a repetition of the sequence $BC$. This repetition may reduce the interestingness of the prediction for the user. Thus, removing repetitive sections improves the quality of the prediction.

**Action Sequences for Adaptive Interactive Systems**

In the following, we discuss how we employ the sequence prediction algorithm in adaptive interactive systems. The approach comprises two steps, training the algorithm and predicting action sequences at runtime. First, rules are extracted either from recorded or live interaction data. For this purpose, a list of user actions must be recognized in the interaction sequence first. In order to generate rules from recorded data, the training algorithm is applied to a sequence of log files. In this case, the training has to be performed only once and is performed offline. Alternatively, the training algorithm collects data from the current interaction. Since training the algorithm after each observed action has a negative impact on the performance of the interactive system, the training is performed either at fixed intervals (e.g. after every fifth interaction) or during idle times of the interactive system. At runtime, the system observes a sequence of user actions, which are for instance detected by the PDFA approach presented in Section 3.2, and finds rule triggers. In the example TV system, if the user often only selects a specific channel in the EPG and then opens the results, the algorithm may extract the sequence ("Go to program guide", "Go to channel selection" / "Selection channel", "Show results"), with the first two actions being the rule trigger. Therefore, if the user opens the channel selection screen, the rule predicts the last two actions of the implication.

Once a sequence prediction has been computed, the adaptive interactive system offers a shortcut to the user. In doing so, the user may execute the action sequence with a single interaction. An example application is presented in Section 6.4.6. The adaptation allows the user to either execute the whole sequence or only a subsequence of the predicted sequence. For example, if the rule of the previously presented example has been triggered, the system offers shortcuts for the user to execute the whole sequence with a single interaction. Thus, once the user executed the first two actions, the remaining actions require only a single click. A second use of the prediction is providing help that explains how to execute the sequence.

However, an adaptive interactive system that employs a prediction of actions requires additional information about the user actions, for example by means of a semantic annotation. First, some actions terminate action sequences, such as actions that open the main menu and thus end a sequence in a submenu. Second, some actions cannot be executed automatically, such as safety checks that always have to be confirmed by the user. Therefore, the system designer has to provide this information to allow the adaptive interactive system to handle actions appropriately.

### 3.4.3 Invoking User Actions

In order to automatically execute action sequences, the system has to trigger the individual actions of a sequence. How to invoke actions depends on the formalism used for detecting user actions. If actions are defined by means of annotated sequences (see Section 3.2), they are played back by simulating these events. However, system reactions are not invoked, because the system has to react to the simulated input in the same way as to actual user input. For example, the simulation invokes a button press, but no system reactions to the user input.

If user actions are connected to events, as discussed in Section 3.2, the action is simulated by invoking the respective event. In any case, the system designer has to consider possible side effects. Therefore, the system may provide shortcuts to the user by playing back the individual actions of a sequence.

## 3.5 Discussion

In this chapter, we introduced a novel approach for describing user actions in adaptive interactive systems. Starting with an observation of basic observations, different algorithms build on each other to produce information that triggers adaptations. First, we discussed the notion of user behavior as a sequence of basic events and the notion of user behavior as actions and data. Basic events encode all aspects of the user-system interaction that are relevant for user modeling. User actions can be extracted from the interaction sequence and tasks describe higher-level user behavior.

We devised a PDFA-based algorithm for extracting user actions from the sequence of events. Based on annotated log data, the algorithm detects user actions, which serve as a basis for a higher-level description of user behavior. We introduced task models as a means for describing higher-level interaction. So far, task models have been used at design time. Instead, we use task model as a means to derive different kinds of information as adaptation triggers, such as user problems or a prediction of the next user action. Different algorithms for prediction both single user actions and sequences of user actions were presented that we adapted to the domain of interactive systems. In addition, we discussed the use of the predictions in adaptive interactive systems.

The algorithms presented in this chapter facilitate a sophisticated description of user behavior in adaptive interactive systems. The approach starts with an observation of basic events and employs different algorithms that work on top of each other to describe higher-level user behavior. This information serves as a trigger for adaptations. In the following chapter, we present usability principles for adaptations in interactive systems as well as a set of adaptation patterns. These use the outcomes of the user modeling process as triggers. In Chapter 6, we prove the feasibility of the user modeling algorithms in an evaluation.

# 4

## Adaptation Patterns for Interactive Systems

To understand is to perceive patterns.

–Isaiah Berlin (1909-1997)

The aim of adaptations is to improve the usability of interactive systems. In the course of this, adaptations have to consider the usability fundamentals of interactive systems. Usability has been a research topic for decades and research and practical experience produced a number of usability principles. These principles define high-level knowledge and fundamental goals of user interfaces. They also apply to adaptive interactive systems, which represent a special kind of interactive system. In this chapter, we introduce usability principles and discuss how to design adaptations such that they comply with these principles. At a first glance, the concept of adaptivity violates some usability principles. For instance, an interface that adapts dynamically to user behavior seems to violate the principle of predictability. Adaptations might cause system reactions that are different from what the user expects. Therefore, we investigate the effects of usability guidelines on the implementation of adaptive interactive systems and discuss how adaptive interfaces comply with usability principles.

Design patterns are a means for communicating best practice. They document general and proven solutions for recurring problems, which are not limited to a specific system or platform. An overview of patterns in human-computer interaction and adaptation patterns in the literature is given in Section 2.3. The origin of patterns is in architecture (Alexander et al., 1977), but patterns have been applied to different domains such as software engineering and interface design. Interface design patterns define best practice of usability in interactive systems and the context in which they should be used. In this chapter, we introduce how adaptation patterns for interactive systems can be identified. We present a format we created for adaptation patterns and introduce a set of patterns we conceived for adaptive interactive systems. In order to provide usable solutions, the adaptation patterns comply with usability principles.

This chapter is structured as follows. First, a set of usability principles for non-adaptive interactive systems is presented and the implications of these principles for adaptations are discussed. Second, the process of identifying and writing down patterns is introduced and a pattern format for adaptation patterns in interactive systems is presented. Finally, we present a set of patterns we created for adaptive interactive systems.

## 4.1 General Considerations for Adaptation Patterns

Adaptations improve the interaction of users with interactive systems. Thus, adaptations should enable the user to interact with the user interface more effectively, efficiently, and in a more satisfying way. However, adaptivity does not fix interactive systems with usability deficits. Instead, adaptations improve the usability of interactive systems that already comply with basic usability principles. In the following, we first introduce usability principles for interactive system and then proceed to discuss the implications of these principles for adaptive interactive systems. Usability experts, researchers, or standards committees have proposed different sets of usability principles in the past. For instance, Shneiderman and Plaisant (2004) present "Eight golden rules for interface design", such as "Strive for consistency" or "Keep it simple". The international EN ISO 9241-110 standard ("Grundsätze der Dialoggestaltung", principles of dialog design; Europäisches Kommittee für Normung, 2006) presents a more formal approach of usability guidelines for dialog systems. These collections all include the usability principles of predictability, transparency, consistency, and controllability, in one form or another.

Since adaptive interactive systems are a special type of interactive systems, usability principles for interactive systems have to be considered in adaptive interfaces. Adaptivity seems to object some usability principles. For instance, a dynamic adaptation of the interface seems to contradict the principle of predictability, because the interactive system might react differently after the adaptation has been applied. However, adaptations may be designed in a way so as to comply with usability principles. In addition to the usability principles of regular user interfaces, usability principles have been discussed specifically for adaptive interfaces.

Literature describes the implications of usability principles on adaptive user interfaces. Jameson (2003) presents a set of usability requirements for adaptive interactive systems and extends these requirements of regular user interfaces for this purpose. In addition, Jameson proposes preventive and compensatory measures that influence usability goals in adaptive interactive systems. In a similar way, Höök (2000) discusses usability issues that prevent a widespread use of adaptations in interactive systems and presents solutions for resolving these problems. Höök mentions control transparency and predictability, privacy and trust, and treating systems as fellow beings. We include the contributions of Jameson and Höök in the discussion of the respective usability principles in the following. In the remainder of this section, we present usability principles for interactive systems, namely "Component Emphasis", "List Element Selection", "Alternative Elements", "Adaptive Help Presentation", and "Shortcut Area". We proceed to discuss the impact of these principles on adaptations.

### 4.1.1 Predictability

A predictable user interface enables users to anticipate the reactions of the system to their input. Users interact more quickly with an interactive system if they do not have to evaluate the system reaction. On the one hand, users learn to use interactive systems automatically if the interactive system reacts in a uniform way to an action, called automatic processing. If users know the reaction, they proceed with the next interaction step without evaluating them first. If an adaptation changes the layout or dialog flow of an application, automatic processing ceases to work and thus deprives expert users of an efficient means of interaction. Therefore, adaptations should not break automatic processing. On the other hand, consistency contributes to predictability by allowing users to anticipate the reaction of the system even in previously unused parts of the system. For instance, if the graphical layout or the speech commands in conformation screens share a common layout in all parts of an interactive system, users know how to operate all confirmation screens. Therefore, once users understand the basic concepts behind the interaction, they are able to operate all parts of the interactive system. Self-explaining and consistent usage concepts contribute to predictability in interactive systems.

The principle of predictability plays a crucial role in adaptive interactive systems. Adaptations should not break the predictability of the interactive system. In addition, adaptations should be implemented in a predictable way. First, the selection of adaptations should be performed in a way so as to enable the user to anticipate which adaptation is selected. If the user shows a specific behavior, the adaptive system should always apply the same adaptation. Second, if an adaptation recommends an element in a list or an action to the user, the selection of the value should be comprehensible and consistent.

The prediction of values by user modeling algorithms should not surprise a novice user and at the same time should be predictable for expert users.

Some adaptations might still change the system in a way that does not comply with predictability and consistency, for instance if the layout of a part of the user interface is to be changed. In this case, the interface may be split into two parts, a non-adaptive, fully predictable part and an adaptive, not fully predictable part. An example of such an interface is the "Shortcut Area" adaptation pattern presented in this chapter.

### 4.1.2 Transparency

Transparency allows the user to comprehend the inner workings and the current interaction state. In doing so, it enables the user to build a mental model of the interactive system. The user can understand the reactions of the interactive system and the system does not confuse the user through unexpected reactions. In addition, the user may better plan future actions, especially longer action sequences that depend on the state of the interactive system. However, the interactive system does not necessarily communicate the actual inner workings and state. Instead, the user sees a simplified model of the interactive system, which does not contain unnecessary details or technical aspects. Höök (2000) calls this approach the "glass box" approach. The inner workings of the adaptive interactive system are a black box, which hides details from the user. However, the interactive system presents a "glass box" to the user that shows a simplified model and allows the user to build a mental model for the interaction more easily.

Explaining adaptation and user modeling decisions to the user is an important aspect for transparency. The interactive system may use different methods to communicate these decisions to the user, such as explicit information. On the other hand, the interactive system may allow the user to build a mental model of the adaptation decision and thus dispenses with the need to explain the decision. In any case, the adaptation decision is an important aspect for transparency. The literature discusses different approaches for transparency in adaptive interactive systems. For instance, Herlocker et al. (2000) present an approach for explaining recommendations in collaborative filtering systems.

### 4.1.3 Controllability

Controllability postulates that the user should keep control over the interactive system and the interaction, including the direction and speed of the dialog. Moreover, the user should be able to control behavior and appearance of the system, for instance by means of configurability. Controllability ensures that the user keeps the feeling of being in control and not being controlled by the interactive system.

System-initiated adaptation performs changes to the interactive system that are not controlled by the user and thus apparently contradict the principle of controllability. Therefore, adaptations should be performed in a way so as to let the user keep the feeling of control. In addition, further methods enable the user to keep control over the interactive system. For instance, the interactive system may allow the user to disable or enable adaptations globally or on a more fine-grained level, e.g. in a settings menu. On the level of individual adaptations, the system may allow the user to approve adaptations before they are performed or allow the user to correct adaptation decisions, e.g. by reverting them. Both methods however require a rich and complex user interface that possibly confuses beginners.

Some user modeling systems allow the user to inspect and alter the user model. For instance, Cook and Kay (1994) present a user modeling server that allows users to browse the user model, view the individual entries, and change the entries according to their preferences. However, users are reluctant or unwilling to perform changes themselves unless the users judge the effort as rewarding (cf. Mackay, 1991). Therefore, different approaches exist that allow the user to keep control in adaptive interactive systems.

### 4.1.4 Unobtrusiveness

Unobtrusiveness is another principle that is of great relevance for adaptive interactive systems. The user interacts with an interactive system in order to accomplish a specific goal and the system should distract or obstruct him or her as little as possible. Since adaptations change the system at runtime, they possibly distract the user. Thus, adaptations may be limited to a specific part of the interactive system. Some adaptive interfaces have been presented in the literature that failed because they were too obtrusive. For example, users regard anthropomorphic agents as too intrusive (cf. Jameson, 2003). Therefore, adaptations should be implemented in an unobtrusive and non-distracting way.

However, adaptations have to be implemented such that the user notices them. During the development of a prototype that implements the "Adaptive help" adaptation (see Section 6.4.2), we observed that some users did not perceive the help messages. Therefore, we increased the degree of obtrusiveness such that the users became aware of the adaptive help, without interrupting the user-system interaction. Therefore, adaptations should be implemented in a way that they do not distract the user from the current task, but have to be intrusive enough to be perceived by the user.

### 4.1.5 Privacy

Although privacy is not a usability principle, it nevertheless represents a highly relevant fundamental for adaptive interactive systems. Adaptation to user behavior relies on an observation of the user-system interaction and the interactive system stores a user model with sensitive data that represents the user.

While users might not be concerned by the fact that the interactive system observes them and stores data (cf. Section 6.2), the user model still needs to protect sensible user data from unauthorized access. On the other hand, sharing data across different applications increases the utility of a user model. Data does not have to be collected from scratch by each adaptive system. If data from the user model should be shared, the user modeling component has to allow the user to decide which parts of the user model to share and which applications may access the user model.

Kobsa (2007) discusses different methods for enabling privacy in personalized interactive systems. According to Kobsa, users want to know which data is collected and want to be able to control how this data is used. Solutions for ensuring privacy in adaptive interactive systems are for example anonymous user modeling or client-side personalization, in which user data is not stored on a public server. In addition, the interactive system may provide a private mode that allows the user to disable user modeling on a per session basis.

## 4.2 Creating Patterns

Patterns represent proven solutions for frequently occurring problems in a specific domain. For this purpose, experts in a domain have to identify problems and solutions that are worth being documented as patterns. This includes problems that are common to different instances and generic. The process of creating patterns comprises two stages: collecting patterns for a domain and writing them down. For this purpose, a common structure is used for all patterns in a pattern collection. In the following, we discuss the process of identifying and collecting patterns. Thereafter, we present a pattern format we defined for adaptation patterns in interactive systems.

### 4.2.1 Identifying Patterns

The pattern author is an expert in the respective domain, such as architecture, software engineering, or usability. He or she wants to share experience and knowledge with other people working in the same domain. As an expert in the domain, the pattern author observes frequently occurring problems and collects successful solutions for these problems. When a solution was applied often or observed in other work or in the literature, the problem and the solution are documented as a pattern. A discussion of the context further defines the pattern.

Contrary to patterns, interface design guidelines define general rules, such as layout, appearance, and behavior of a specific platform, rather than generic knowledge for arbitrary interactive systems. The Apple Human Interface Guidelines (Apple Inc., 2010) represent one example of interface guidelines. These define a common look and feel for applications on the Mac OS software platform. A closed group, such as a platform vendor or a committee, often defines guidelines. Unlike guidelines, design patterns describe knowledge that has proven successful in practice and was discovered empirically. Pattern collections show varying degrees of generality. On the one hand, the user interface pattern collection presented by Tidwell (2005) comprises a large number of patterns, some of them discussing very specific problems. On the other hand, the pattern collection for adaptive hypertext systems presented by Koch and Rossi (2002) includes a smaller number of patterns that discuss more general problems and solutions.

Once recurring problems have been identified in a domain, the pattern author defines a pattern format for the pattern collection and writes the patterns down in the common format. In the following, we introduce different pattern formats from the literature and present a format we defined for the adaptation patterns in this work.

### 4.2.2 A Pattern Format for Adaptation Patterns for Interactive Systems

Pattern formats have been defined for various domains. The original pattern format is the Alexandrian form. It was introduced by Alexander's pattern language for architecture (Alexander et al., 1977). The Alexandrian form, which is discussed in detail in Section 2.3, divides the pattern description into two sections, a discussion of the problem and a discussion of the solution. Unlike later formats, the Alexandrian form does not use explicit headings. Another widely used pattern format is the "gang of four" format for software engineering patterns by Gamma et al. (1995). It divides a pattern description into sections with explicit headings. All patterns use common sections, such as "Pattern Name and Classification", "Intent", or "Implementation". These sections enable readers to find interesting sections and skip uninteresting ones.

The pattern format by Gamma et al. (1995) serves as a basis for the adaptation patterns presented in this work. However, we removed inapplicable sections, such as "Implementation" or "Sample code", and added a section "Adaptation trigger" to include the information how an adaptation is initiated. In the following, we introduce the individual sections of the adaptation pattern format.

- **Name**: An intuitive name allows the developer to easily identify and remember adaptation patterns.
- **Intent**: The intent describes the idea behind the pattern in a concise statement. In doing so, the intent sketches the solution briefly.
- **Motivation**: The motivation describes the problem that is addressed by the pattern and introduces the motivation for the presented solution.
- **Forces**: The forces section further defines the context of the pattern by listing pieces of information that are relevant for the application of the pattern. For instance, forces include general conditions or limitations of the pattern.

- **Solution**: The solution presents a method for solving the previously described problem. In addition, this section discusses the merits of the adaptation.
- **Adaptation Trigger**: The adaptation trigger connects the pattern to the user modeling and discusses user behavior that activates the adaptation. For this purpose, this section presents an enumeration of user modeling entities, such as user preferences or a prediction of user actions.
- **Related Patterns**: The context of the pattern is further defined by connecting the pattern to others, both from the same and from other pattern collections. Moreover, if the pattern does not solve the problem of the developer, this section suggests other patterns that could be applied instead.
- **Examples**: A number of examples further illustrates the use of the patterns. Screenshots of adaptive applications clarify the use of the pattern.

In the following, we present a number of adaptation patterns we defined for interactive systems. The full description of these patterns in the format introduced in this paragraph is presented in Appendix A.

## 4.3 An Adaptation Pattern Collection for Interactive Systems

In this section, we present a set of adaptation patterns for interactive systems. These change the user interface of the system, but do not depend on the application logic. Since some adaptation address specific issues rather than general problems, they are outside of the scope of this work. For example, such adaptations include an adaptive route generation algorithm of a navigation device by the Adaptive Route Adviser (Langley et al., 1999). Instead, this work deals with generic adaptations for user interfaces. These patterns discuss the adaptation of multimodal interactive systems by considering both graphical and speech interfaces. The patterns presented in this section are more general in nature than interaction pattern collections (e.g. Tidwell, 2005) and correspond in their generality to the adaptation patterns for hypertext systems by Koch and Rossi (2002). The adaptation patterns consider the usability principles presented in Section 4.1.

Observations of the user-system interaction trigger adaptations. For this purpose, a user modeling component creates a representation of the user from these observations. This representation serves as a basis for adaptation decisions. The model uses different algorithms such as the ones presented in Chapter 3. The modeling process is not part of the pattern descriptions. Instead, the "Adaptation Trigger" section of the patterns describes observations of the user modeling component that activate the adaptations.

In the following, we present a set of adaptation patterns we devised for interactive systems. This section summarizes the patterns. A detailed discussion that uses the pattern format introduced in Section 4.2 is presented in Appendix A. We identified these patterns by reviewing adaptations in the literature and classifying adaptations used in prototype systems we developed for this work.

### 4.3.1 Component Emphasis

A user performs actions to achieve a specific goal. After user modeling algorithms have recognized this goal, the "Component Emphasis" guides the user to it. For this purpose, the system emphasizes interface elements that are connected to predicted or recommended actions. The user employs interface elements to control the system and the elements are connected to specific user actions. For example, a graphical button or a speech command may open the EPG in a TV system. During the interaction, the user may spend considerable time looking for a specific element or may not know which action to use next. The adaptation changes properties of the elements in a way as to draw the user's attention to them. Thus, emphasizing the respective elements guides the user to the associated actions. In doing so, the user finishes the current task more quickly or gets to know actions that have not been used before.

Different assumptions of the user modeling component may trigger the adaptation. This includes either a prediction of the next action or a recommendation of the user modeling component, for instance for actions the user has not used yet. Since an emphasis of wrong elements impedes the user, an appropriate user modeling prediction is crucial for this adaptation. The adaptation only performs small changes of the interface by emphasizing elements. Significant changes of the interface may either distract a user from the task or hinder the user in working with the interface as he or she knows it. Therefore, subtle guidance supports the user. The adaptive emphasis should be performed in a way such that the user does not confuse it with a regular selection in the user interface. The changes are limited to the part of the interface that requires emphasis. In doing so, users are enabled to reuse acquired knowledge of the interactive system. Thus, distraction of the user through fundamental changes of the interface is avoided.

A graphical button may for instance be emphasized by increasing its size or selecting more noticeable colors. The emphasis may also be visualized by means of an animation. In both cases, the user notices the button he or she is looking for or knows which action to perform next. In a speech interface, the interactive system may read a list of possible commands for each context. Putting them to the beginning or end of the list emphasizes individual commands. An example shall illustrate the adaptation. In an EPG, the user specifies filter criteria, such as channel or time, to filter the list of TV shows. After a number of criteria were selected, the user has to press a "Show results" button to see all shows that match the selected criteria. Increasing the size of the button and changing colors emphasizes the button. In doing so, the user may finish the task more quickly. The adaptation may also be applied to voice interfaces. If a user enters the filter selection screen, the system may read out the command for opening the result list as the last element. This pattern is presented in the adaptation pattern form in Appendix A.1.

### 4.3.2 List Element Selection

The selection of elements from a list is a frequent task in interactive systems. For example, a user may select names from an address book or a city from a list in a navigation device. However, users select specific entries more frequently than others. For instance, a user calls close friends more often than other contacts. An adaptation may improve this task by enabling the user to select frequently used entries more quickly. For this purpose, the List Element Selection adaptation emphasizes specific entries in a list. In doing so, the adaptation enables the user to see at a glance which list entries are more relevant than others.

The user modeling component identifies list entries that are interesting for the user. For this purpose, it selects entries that are selected more often than the rest, either by an individual user or by a user group. In addition, elements may be chosen that have not yet been selected, but are deemed to be interesting according to the user's behavior. Since an emphasis of wrong elements impedes the user, the prediction quality is of great importance. The reasoning behind this adaptation is that the selection of frequently used items in a list should take less time than the selection of others. Emphasized list elements should be highlighted in a way such that the user does not confuse the emphasis with other markings, such as a selection cursor.

The interactive system may highlight entries by selecting more noticeable text or background colors, adding visual markings, or increasing their size. For instance, selecting a name from the address book is one of the fundamental functions of interactive systems that support phone calls, such as mobile phones or automotive dashboard systems. Since users call a small number of people from their phone book frequently, highlighting the names supports their selection. For this purpose, the adaptation selects a more noticeable background color for highlighted elements and adds a special icon to recommended entries. If a list is longer than one screen, interesting items may also be marked in a scrollbar to enable the user to see highlights in the full list. Speech interfaces emphasize list elements by reading interesting elements first or by adding acoustic markings. Whereas the "Component Emphasis" adaptation emphasizes interface elements that trigger actions, the "List Element Selection" highlights one item in a list of similar elements. In Appendix A.2, we present the description of this adaptation in the pattern form.

### 4.3.3 Alternative Elements

Users of interactive systems differ with regard to preferences and characteristics, such as the proficiency with computers or eyesight. Therefore, a single interface, however well designed it may be, does not reflect the individual requirements of a user in an optimal way. Instead of providing one configuration that tries to consider all users, the "Alternative Elements" adaptation enables an interactive system to select the most appropriate version of an interface element. For this purpose, the system designer provides different alternatives for interface elements, such as graphical screens or speech output prompts. In addition, the developer specifies the characteristics of the individual elements. For instance, this may be information that is element is designed for beginners or experts. Based on this information, the adaptation in turn selects the most appropriate one.

A user modeling component computes information about characteristics of the user that enables the adaptation to select one among the alternatives. This information includes preferences or characteristics of the user, such as the proficiency with an interactive system, the age, or impairments. By providing a set of alternatives to the adaptation that were created by the system designer, all variants of the interactive system take into account design principles. Automatically generated alternatives can break with existing usability principles. However, the developer has to spend additional time creating the alternatives, but the user benefits from an improved usability. Significant changes should be communicated to the user in order to avoid confusion, for instance with a notification.

This adaptation may be applied to different parts of the user interface and at different levels. In graphical interfaces, the adaptation may be applied to elements ranging from complete graphical screens to individual items. In speech interfaces, it may choose among different sets of speech output prompts. In addition, the adaptation may select among different dialog flows and system properties. For example, the adaptation selects among different versions of a navigation destination input screen. A simple version is provided to novice users and a more powerful one to advanced users. On a lower level, a larger font size is chosen to improve the readability for visually impaired users. A speech interface may provide different versions of speech output prompts. Novice users receive extended prompts that explain the most important functions. Intermediate users only require shorter prompts, which list the commands. Finally, expert users, who could be annoyed by long and repetitive speech output, only hear a short prompt explaining the current state of the system. This pattern is presented in the adaptation pattern form in Appendix A.3.

### 4.3.4 Adaptive Help Presentation

The user may access online help in interactive systems faster than printed documentation. However, this help is often static or only considers the current screen. People are likely to have different problems and needs. For instance, beginners need general and extensive help, whereas experts seek specific solutions and might be annoyed by superficial help. Adaptive help considers the user's current situation and background. For this purpose, it takes into account not only the current context, but also other information, such as the user's interaction history and characteristics. In doing so, help is more apt and thus supports a user more precisely in the current task.

The user modeling component observes the user-system interaction to determine the situation and characteristics of the user. Adaptive help may be provided for a prediction of the next user action. In addition, the system may detect user problems, such as hesitation or aimlessness. When deciding about the selection of help, the system takes into account preferences and characteristics of the user, such as the knowledge level in the current context or general experience with the system. Adaptive help should not surprise, annoy, or obstruct the user. For instance, a user notices a help dialog that has to be closed explicitly. However, it may interrupt the user's train of thought and thus distract from the task. The help messages have to be precise, since users do not read these messages otherwise. At the same time, they have to offer enough new information that users regard them as helpful. Help may be assistive for beginners, but annoying for expert users.

Instead of overlapping the interface, the help messages may be presented on a separate area of the screen. Alternatively, an icon (or a sound) may indicate the availability of help that the user opens explicitly. The messages should not fully engage the user's attention, as for instance modal help messages do. In speech interfaces, an acoustic signal may be used instead of a graphical hint. An example shall illustrate the "Adaptive Help" adaptation. In an interactive TV system, the user browses the TV program in an EPG with different filter criteria, such as channel or time. Help is presented by fading in a yellow message box on the top of the screen. When the user enters the selection screen for the first time, the help explains how to select filter criteria. After some criteria were selected, the help text on the screen tells the user to open the result screen next. For this purpose, the interaction history is processed. In Appendix A.4, we present the description of this pattern in the adaptation pattern form.

### 4.3.5 Shortcut Area

Users perform specific actions or action sequences repeatedly. For example, users apply settings after they start a jointly used system or select specific elements in a list. The "Shortcut Area" adaptation offers shortcuts to actions or data by presenting them in a separate area of the screen, called shortcut area. In doing so, the adaptation does not interfere with the interaction and enables the user to decide whether to use the shortcuts. Thus, the adaptation does not distract the user. The adaptation may provide two kinds of shortcuts. First, the shortcuts select among different alternatives. For example, in an address book, the adaptation selects the most frequently used names and places them in a separate area on the top of the list. Second, the shortcuts may represent a sequence and allow the user to select a subsequence. For instance, if the shortcuts represent a sequence of actions, the user selects an action and the interactive system executes the sequence from the first one to the selected action.

The user modeling component computes the list of shortcuts. Based on an observation of user actions, it provides a list of alternative actions or a sequence of actions. Alternatively, a list of preferences may be created, such as a list of the user's favorite TV channels. Shortcuts that automatically appear and overlap the interface distract the user. A separate area that is always visible instead allows the user to decide whether or not to use shortcuts. In doing so, it limits the distraction of the user.

The shortcut area may either be part of one interface element, such as a list, or a separate part of the whole screen for presenting global shortcuts. In both cases, a separate area is reserved for the adaptation. In a selection list, a separate area on the top of the list presents the most frequently selected entries of the list to the user. By selecting them, the user does not have to scroll through the whole list. A different application of the Shortcut Area pattern is to provide navigation shortcuts. Based on a prediction of a sequence of actions, each action is represented by a button. If the user presses one of these buttons, the action associated with the button and all actions before the pressed one are executed. In doing so, the user may reduce the number of interactions. The pattern is presented in the adaptation pattern form in Appendix A.5.

## 4.4 Discussion

In this chapter, we introduced usability principles for interactive systems, such as transparency or controllability, and the implications of these principles for adaptations in interactive systems. We presented patterns as an approach for sharing knowledge and best practice in a domain. The process of identifying and defining patterns was discussed. Moreover, we defined a format for adaptation patterns in interactive systems. We devised a set of five adaptation patterns, which are discussed in more detail using the previously defined pattern format in Appendix A. These patterns include references to other patterns, both from this pattern collection and pattern collections in the literature. A number of examples both from graphical and speech-based interactive systems illustrate the patterns. These patterns support a developer in selecting successful adaptations. In doing so, adaptive features may be integrated more easily into any interactive systems. In the following chapter, we present an adaptation framework that includes these patterns. In Chapter 6, we discuss an evaluation of interactive systems that use these adaptations.

Whereas the pattern collection covers all adaptations that we identified during the pattern discovery phase, additional patterns might be added in the future. The pattern collection may evolve into a pattern language, which offers a comprehensive description of patterns in a domain. Our adaptation pattern collection already fulfills many requirements of a pattern language, such as a common naming convention and interconnectedness.

# 5

## AdaGUIDE – An Adaptation Framework

Vision without implementation is hallucination.

–Benjamin Franklin (1706–1790)

In addition to usability and the selection of useful adaptations, Höök (2000) identifies development methods as a crucial prerequisite for a widespread adoption of adaptive interfaces. We presented an approach for user modeling from basic events in Chapter 3 and a set of adaptation patterns for interactive systems in Chapter 4. On this basis, we devised an adaptation framework as part of this book to allow a developer to integrate the presented approaches into interactive systems more easily and conveniently. This generic adaptation framework provides a reusable platform for adaptive interactive system and improves the task of developing these systems. For this purpose, it integrates user modeling and adaptations. First, using an existing framework reduces the development time, because the developer does not have to implement all the components of the framework. In addition, this framework comprises a number of user modeling algorithms and adaptations, which thus do not have to be implemented. Second, the developer may avoid common pitfalls of adaptation architectures by employing an existing architecture rather than developing a new framework. In addition to supporting developers in creating adaptive interfaces, the framework also serves as a test bed for an evaluation

of the adaptation approach. In doing so, the framework shows the feasibility
of our adaptation approach.

In this chapter, we present an adaptation framework for adaptive interactive systems. The main components of the framework are a user modeling
component, an adaptation component, and a semantic layer. The user modeling component observes the user-system interaction and creates an abstract
representation of the user. The adaptation component applies adaptations to
the interactive system. For this purpose, it employs information stored in a
semantic layer that serves as an abstraction of the interactive system. We
created a reference implementation of this framework as an extension to a
model-based development tool. The implementation includes the user modeling algorithms presented in Chapter 3 and the adaptations from Chapter 4.
The reference implementation also serves as a test bed for an evaluation and
verification of these approaches. We show the feasibility of the framework by
implementing adaptive features in different interactive systems and performing user tests with these interactive systems. These systems include a digital
TV system and an automotive dashboard system. Chapter 6 presents an evaluation of the user modeling algorithms and the adaptive test systems, which
employ our adaptation framework.

This chapter is structured as follows. After an overview of the architecture
of the framework, a semantic layer that serves as an abstraction of the interactive system is introduced. Thereafter, a user modeling component is presented
that extracts information from the user-system interaction and computes further derivations with user modeling algorithms. Finally, the integration of
adaptation patterns into the adaptation framework is discussed and a reference implementation of the framework is presented.

## 5.1 Overview of the Architecture

In the following, we present an adaptation framework called AdaGUIDE that
integrates the concepts presented in previous chapters. The aim of the framework is to provide a reusable platform for adaptive interactive systems and to
create a test bed for an evaluation and assessment of the adaptation approach
we present in this work. Figure 5.1 gives an overview of the framework. A semantic layer, a user modeling component, and an adaptation component constitute the main components. The framework extends an interactive system.
Triggered by user behavior, the user modeling component uses algorithms,
such as the ones presented in Chapter 3, to model the user. An adaptation
component integrates the adaptation patterns from Chapter 4 to adapt the
interactive system.

A semantic layer creates an abstraction of the interactive system and other
aspects relevant for adaptations. It serves two purposes. First, this layer creates a uniform representation of different components, such as the interactive

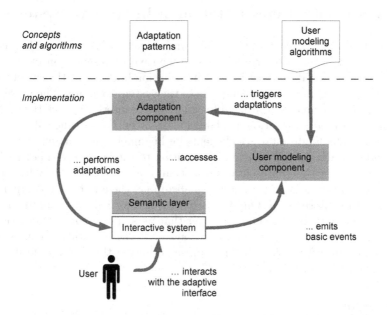

**Fig. 5.1.** The architecture of the adaptation framework comprises a semantic layer that creates an abstract representation of the interactive system. A user modeling component observes the user-system interaction and triggers and an adaptation component.

system and the user model. In doing so, the semantic layer provides uniform access to different aspects of the interactive system for the adaptation component. Second, the semantic layer serves as an abstraction for using the adaptation framework with different interactive systems. The semantic layer employs semantic technologies and allows reasoning on the data.

A user modeling component observes the interaction by watching events from the interactive system. This component serves two purposes, namely collecting information about the user and employing algorithms to derive more information. The user modeling component may use the algorithms presented in Chapter 3, but the generic structure allows other user modeling algorithms to be implemented as well (see Zukerman and Albrecht, 2001). Based on the information in the semantic layer and the outcomes of the user modeling process, an adaptation component decides about adaptations to better reflect the characteristics of an individual user. For this purpose, the adaptation component employs a formalized description of the adaptation patterns presented in Chapter 4. The following sections discuss the individual components in detail.

## 5.2 A Semantic Representation of Interactive Systems

In order to decide about adaptations for an interactive system, an adaptation component requires a uniform and comprehensive representation of the system. This representation must include information both about the system and the user to decide how and which parts of the system to adapt. The representation of this information is crucial for an adaptive interface, since the system may only consider information available in this layer. For instance, if an adaptation should highlight interface elements for beginners, the semantic layer has to include proficiency information. In addition to the amount of information, the formalism for this representation plays an important role. Only by representing all aspects using a common formalism may the adaptation component access all this information. Ontologies represent a formalism for representing a domain. We introduce a semantic layer that uses an ontology for describing the domain of the adaptive interactive system. The adaptation component may exploit all information that is represented using the common formalism in the adaptation decision.

### 5.2.1 Ontologies

An ontology is a conceptualization of a domain in a machine-understandable format. Ontologies, which have their roots in knowledge management (Gruber, 1995; Noy and McGuinness, 2001), define a domain by a set of entities and the relationships between these entities. A specific type and a set of properties define the entities. Properties have either primitive data types, such as string or number, or are connections to other entities. Ontologies consist of statements or triples, which comprise a subject, a predicate, and an object. The subject is an entity from the ontology, the predicate corresponds to a property, and an object is either an ontology element or primitive data. Statements are visualized either with text or using a graphical notation. The example in Figure 5.2 presents the graphical notation of statements of a wine ontology. A set of types defines the templates for the ontology, in this case "Wine" and "Estate". The "Wine" type includes properties such as grape, year, and wine estate. Individuals instantiate types to define specific entities, such as a specific wine (e.g. Chateau Margaux, Cabernet Sauvignon, 1983). In addition to storing data in statements, ontologies infer new information from existing information, e.g. by means of inference rules. Ontologies represent a proven approach for storing and accessing complex data.

Different formalisms for ontologies have been defined, such as DAML+OIL and RDF/OWL (see Horrocks et al., 2003). The Web Ontology Language (OWL; Smith et al., 2004) extends the Resource Description Framework (RDF) with a formalism for defining types, called classes. Thus, items of the same type are represented formally by a common class and share the same set of properties. Inheritance defines a hierarchy of classes that inherit properties from their parent class. Individuals instantiate classes and values are assigned

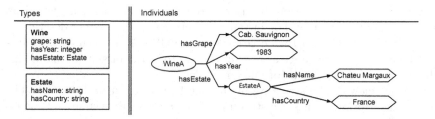

**Fig. 5.2.** An example of a wine ontology, which is represented by means of a graphical notation.

to the properties of the respective class. Inference may be performed on the OWL information, such as inheritance information. Restrictions add further conditions to properties that must be valid. We use the OWL notation in this work for different reasons. First, OWL enables a reuse of existing domain ontologies. Since OWL is an accepted standard, many domain ontologies have been defined using OWL. For instance, the OntoSelect repository (Buitelaar, 2004) or the SchemaWeb directory[1] provide repositories of OWL ontologies. In addition, existing OWL tooling may be used, such as the widely used graphical ontology editor Protégé-OWL (Knublauch et al., 2004). Moreover, different libraries exist that support OWL, such as the Jena framework[2] for building Semantic Web applications, different inference engines (e.g. KAON2[3] or FaCT++[4]), or querying languages such as SPARQL (Prud'hommeaux and Seaborne, 2008). By using the OWL notation, all these technologies and existing libraries may be reused.

An OWL ontology forms the basis of the semantic layer in this framework. This ontology covers different aspects of the adaptive interactive system, which are described by a set of OWL classes. In the following, we discuss the individual parts of the ontology for adaptive interactive systems.

## 5.2.2 An Ontology for Interactive Adaptive Systems

In this framework, we employ an ontology to describe different aspects of the interactive system, such as the system itself, the user, the user-system interaction, and adaptations. The ontology provides a uniform access to different kinds of information. An adaptation component reads and updates this information. The ontology consists of a set of OWL classes with a set of properties. For example, the name of a user represented by the User class is stored in a property called hasName. The classes form a hierarchy with a common base class called DialogThing. Each part of the ontology, such as the description

---

[1] SchemaWeb: http://www.schemaweb.info/

[2] The Jena framework: http://jena.sourceforge.net/

[3] KAON2: http://kaon2.semanticweb.org/

[4] FaCT++: http://code.google.com/p/factplusplus/

of the interactive system or the user-system interaction, is again based on a common ontology class. For instance, all classes of the system description are derived from the `DialogSystem` class. The OWL formalism encodes further knowledge, for instance as restrictions of properties or general knowledge about the system by means of inference rules.

An overview of the structure of the ontology is given in Table 5.1. The system description defines the structure of the interactive system and the interface elements, such as graphical buttons and lists or speech components, as well as their properties. The dialog domain describes the domain of the interactive system, for instance a list of channels in an interactive TV system. The domain description may incorporate existing ontologies. User modeling information reflects data specific to the current user. In addition, the user system interaction is described by actions and tasks. Finally, the ontology includes a description of adaptations. The ontology is restricted to a declarative description of the system and describes the structure rather than the workings of the system. For instance, the ontology does not include information about the specific appearance of interface elements, but is limited to aspects that are relevant for the adaptations.

### The System Model

The system model provides classes for a technical description of the interactive system and includes a description of the graphical and speech components. In addition to the technical description, which may be generated from an existing system description during the development phase, semantic information is annotated to the system model to further enhance the scope of the knowledge base. In doing so, these annotations may connect the system model with domain information. For example, all elements, such as graphical buttons or speech output prompts, may have a "type" property to describe their purpose, e.g. "Help".

As an example formalism, the implementation of this adaptation framework (see Section 5.5) supports a statechart-based formalism (Harel, 1987). Statecharts consist of states and transitions between these states. Transitions are triggered by events. Therefore, the system model provides classes for states, transitions, events, etc. However, interactive systems that rely on a different formalism, such as USIXML (Limbourg et al., 2004) or XUL[5], can be represented by a different set of classes. The system model presents an abstraction of the system for the adaptation layer, which reads information about the system and performs changes through this representation.

### The User Model

The user model describes preferences and characteristics of different users of the interactive system. On the one hand, this model comprises information

---

[5] XML User Interface Language: http://www.mozilla.org/projects/xul/

| Model | Describes ... |
|---|---|
| DialogSystem | ... the dialog system on a technical level, e.g. graphical elements and speech components |
| DialogDomain | ... the dialog domain |
| UserModelItem | ... users and entries from the user model |
| Interaction | ... user actions and tasks |
| Adaptation | ... adaptations and to which elements they can be applied to |

**Table 5.1.** The structure of the ontology with the five sub-models. All models are derived from the OWL base class Thing and a common base class DialogThing.

about the user, such as name or preferences. On the other hand, it includes technical information, such as the configuration of the system for a user, for instance the state of interface elements.

However, not all user-related data is stored directly in the knowledge base, since the OWL formalism does not support complex data types well. For instance, storing the data of a matrix in the OWL notation is complex and inefficient, requiring a significant number of statements to represent the individual rows and columns. Instead, a separate user model stores arbitrary complex data types. In order to include data from the user modeling component in the knowledge base, a bridging component synchronizes selected data. A common type system of the knowledge base and the user modeling component enables a direct data conversion.

### The Interaction Model

The description of the user-system interaction plays a fundamental role when adapting interactive systems to user behavior. The interaction model describes user actions and is therefore an essential component of the framework. Different elements define interaction at several levels of abstraction. At the lowest level, basic events represent observations, such as a key press or an internal state change of the system as a reaction to the key press. User actions consist of sequences of these basic events and define logical steps in the user-system interaction. Tasks describe the user-system interaction at a higher level and are hierarchical constructs composed of user actions. The ontology does not cover the internal structure, but only the description of the elements. For instance, information about the composition of user actions out of low-level events is not part of the semantic layer, but only the list of these actions.

### The Adaptation Model

The adaptation model defines adaptations for interactive systems. Each one consists of a name and a definition of which elements it can be applied to. However, this model does not define the execution of the adaptations. Instead, a separate component specifies the execution at a functional level. For instance, an adaptation called "Component Emphasis", which highlights a graphical element in the interface, consists of the name "Component Emphasis" and a list of interface element types it can adapt, e.g. graphical buttons.

## The Domain Model

The domain model covers domain- and application-specific knowledge. Since it depends on a specific domain and interactive system, the knowledge can in most cases not be reused. For instance, the domain model of an interactive TV system, which covers information such as the list of TV channels, is not applicable to an automotive dashboard system.

The adaptation component does not require an ontology that compiles extensive knowledge across domains, as for instance the Cyc project (Lenat, 1995) does. Rather, the interactive systems ontology only covers these pieces of information that an adaptation component needs to derive adaptations. Thus, the scope of the domain model depends on the requirements of a specific adaptive interactive system. The domain model may however be connected to existing ontologies, if an appropriate ontology already exists. Existing ontologies can for example be merged into the domain model by means of ontology merging (see Noy and Musen, 2000).

### 5.2.3 Instantiating the Models

In the previous section, we introduced a number of models that describe different aspects of interactive systems by a set of OWL classes. In order to describe one specific system, instances of these classes have to be created. For this purpose, an individual is created for each entity of the interactive system by instantiating the respective classes. A set of ontology connectors produces these individuals and each connector covers a different aspect of the system. For instance, a system model connector creates an individual for each interface element, state, and transition. An interaction model adds an individual for each user action. At startup, the interactive system executes all ontology connectors to generate a common representation of different aspects of the system. Therefore, these connectors bring together data from different information representations into a common ontology-based formalism.

Figure 5.3 gives an example of ontology classes and individuals that are created by ontology connectors. A set of ontology classes provides templates for individuals, which are State, Transition, and Session in this example. The ontology connectors create individuals for the different classes, for instance for all states. For this purpose, an individual of the type State is created for each state from the system description and the properties are initialized with the respective values. In the example, an individual called State_10 is created and the hasName property is initialized with the name of the state "Main Menu". In addition, state State_10 is connected to state State_5 by

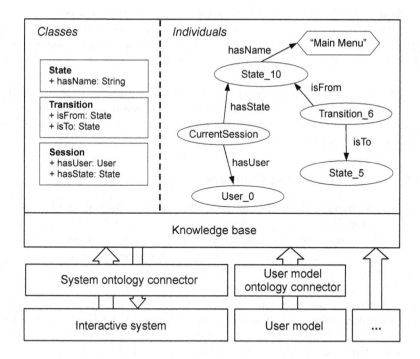

**Fig. 5.3.** Ontology connectors use classes as templates to create an interconnected knowledge base to represent the interactive system.

means of a transition called `Transition_6`. In doing so, the ontology connectors create an interconnected representation of the interactive system based on the ontology classes.

Queries provide access to this semantic representation. A query applies a set of input statements to an ontology and returns all matching ones. For instance, a query may select all individuals for beginners. The adaptation component performs such queries to determine which adaptations to apply. On the other hand, the adaptation component also updates the semantic representation to alter the interactive system. For this purpose, the connectors watch the respective entities and update the interactive system accordingly when entities have been changed.

## 5.3 A Framework for Modeling User Behavior

The requirements of different interactive systems towards a user modeling component vary greatly. For instance, one system may adapt to the user's proficiency, whereas another one may adapt to the user's interaction history.

Therefore, a flexible and reusable user modeling architecture has to support the requirements of different interactive systems. The user modeling component serves two purposes, collecting data from the user-system interaction and using user modeling algorithms to derive information for the adaptation decision.

In this section, we introduce a user modeling architecture. It connects different components that read from and write to a common event bus. Different components communicate through this bus. Finally, a user model applies algorithms to the collected data.

### 5.3.1 The User Modeling Component

A user modeling component forms the backbone of the user modeling framework and provides a generic communication bus for different components. Whereas the component is generic, we present a number of specific user modeling algorithms to illustrate this architecture. An overview of the user modeling component is given in Figure 5.4. The interactive system submits events to the event bus, which are observed by different components, such as an interaction model and a user model. These models perform computations and possibly submit new events. The user modeling component is tightly coupled with the interactive system, but still represents a reusable and independent component that can be employed in different interactive systems.

Events represent the basic interchange format of the user modeling component and consist of a timestamp, a type, and a set of parameters (see Figure 5.5). The timestamp denotes the time at which the event occurs. The event type indicates the event trigger, such as a button press by the user or an internal system reaction to user input. Finally, a set of parameters further describes the event. The parameters depend on the event type. For instance, they contain the name of the button in case of a button press. Different components of the user modeling system exchange events through an event bus, which is conceptually similar to blackboard architectures (cf. Corkill, 1991). Producers submit events to the bus. For instance, the interactive system submits events when it observes a button press, a speech utterance, or a state change. Other components subscribe to the event bus, either to all events or certain event types with specific parameter values. These components are notified when matching events are submitted. Therefore, the event presents a generic interchange format between different components of the user modeling system.

As discussed in Chapter 3, the view of the user modeling process is limited to events the system observes. For this purpose, the interactive system or other sensors emit events to the user modeling component. These events include input from different modalities, such as haptic input, speech, or gesture interaction, as well as system reactions and internal events. Other input types, such as facial expressions detected from a video stream or emotion

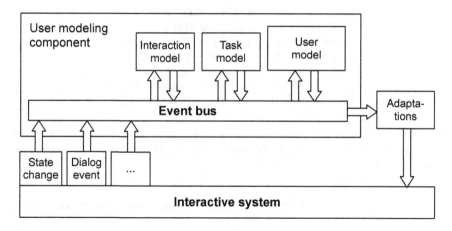

**Fig. 5.4.** The event architecture transmits events from different sources, such as the interactive system or higher-level components.

```
[1234567890123] type param_1={value_1} ... param_n={value_n}
```

**Fig. 5.5.** Structure of interaction events.

recognition from speech input, are beyond the scope of this work. However, the user modeling component may process these if they are encoded as events.

Before the user modeling component processes events, it applies a pre-processing. First, the user modeling component filters out these event types that are not required in the user modeling process, such as internal system reactions. Second, missing information is added to events, such as information about the modality of the events. For example, regular expressions may add properties to events based on the event type and existing properties. Finally, the user modeling component may correct the order of events, if for some reason they do not arrive in the correct order.

Higher-level models observe the events and submit new ones to the event bus. For instance, an interaction model may subscribe to specific events and submit a new interaction event when a certain action was observed. In Chapter 3, we introduce an approach for modeling user behavior in interactive systems. This approach detects user actions from basic events by means of probabilistic automata (see Section 3.2). The user modeling framework includes an implementation of this algorithm. If the algorithm detects a user action, the interaction model generates a new "interaction" event and stores further information, such as the name of the action and a confidence score, as event parameters. Similarly, a task modeling component (see Section 3.3) observes interaction events from the interaction model and detects the task a user is working on. Different tasks are derived from the task model and submitted

```
1 [1180520776220] hw name={COMM_0x6e}
2 [1180520776220] event name={Start}
3 [1180520776220] state name={Session_Main}
4 [1180520776376] view name={WelcomeView}
...
5 [1180520776379] interaction name={Startup}
6 [1180520776381] task name={Welcome} type={started}
```

**Fig. 5.6.** Exemplary log lines, showing a button press (line 1), reactions of the system to this input (lines 2–4), and higher-level events created by the interaction model and task model (lines 5 and 6).

to the event bus. Thus, higher-level models process events to describe user behavior.

Figure 5.6 presents an event sequence from the startup of an interactive system. The user presses a button on an input device (line 1), such as a remote control. The interactive system responds with an internal event called "Start" (line 2), enters a new state (line 3), and opens a graphical screen called "WelcomeView" (line 4). The interaction model observes these events, recognizes the "Startup" action, and emits an "interaction" event accordingly (line 5). The task model in turn observes "interaction" events to detect the task a user is working on. In this example, the task model detects the "Welcome" task and emits an appropriate event (line 6). In doing so, different components of the user modeling framework communicate through generic events.

In addition, a user model is connected to the event bus. This model stores information about the user and computes further derivations. In the following section, we discuss the user model in detail. Finally, an adaptation component receives event notifications and invokes adaptations depending on the outcomes of the user modeling procedure. This sample lineup of components provides a comprehensive description of user behavior. However, additional components may be connected to the event bus. For instance, an interactive system may require specific algorithms for recognizing user actions or tasks. Therefore, a generic event bus connecting different components provides a viable foundation for describing user behavior from low-level events in interactive systems.

### 5.3.2 The User Model

The user model derives user characteristics and preferences from an observation of basic events. In addition to storing user-specific data, the user model serves two purposes: first, extracting information from events and updating user model entries accordingly, and second, computing further inferences by means of user modeling algorithms. We now discuss the responsibilities of the user model in detail. Different entries are connected either to the event bus or other user model entries and the values are updated with the outcomes of user modeling algorithms. Since the semantic knowledge representation introduced in the previous chapter comprises user modeling information, the semantic layer could be used as a data store for the user model. However, specific algorithms require data types that cannot be represented efficiently in an ontology by means of statements. For example, storing a matrix semantically requires a large number of statements, which perform significantly worse than a binary representation. Therefore, we use a separate data store for the user model.

The main purpose of the user model is to serve as a store for user-specific data. This data is stored in user model entries by means of typed name-value pairs, i.e., each entry consists of a type, a name, and a value. The user model supports basic data types, such as strings or numbers, but also more complex types by means of an extension facility. Complex data types store data that is used by specific user modeling algorithms, which is beyond the power of simple data types. Complex data types can be compounds of simple types, such as list, map, or matrix, but may also be arbitrary other types. User-defined data types can be integrated, allowing for very specific data types for algorithms. Figure 5.7 illustrates the workings of the user model. In Figure 5.7.A, an entry "user.name" of the type `string` stores the name of the user "Peter". The user model stores entries separately for each user. Thus, if a different user works with the system, the "user.name" entry contains his or her name. Default values may be defined by means of a template role that is assigned to the user. For instance, the "beginner" role provides values for users who are not experienced with interactive systems. In addition, custom default value generators may be defined to compute specific values for new users, for instance by generating a unique identifier. Restrictions limit the values of a user modeling entry, for instance by defining a range for numbers (e.g. age from 0 to 110 years) or providing a list of allowed values.

The user model also provides facilities for loading and updating values directly from events and values of other entries. For this process called derivation, each user model entry may have a source that defines how the data is loaded or updated. A user model source consists of data elements, which are either events or other user model entries, and an operator that defines how to compute the value. An operator performs a specific computation with the data sources to produce a new value. The simplest operation is to copy the value from the source to the target entry. An example is presented in Figure 5.7.B: the source `Copy` copies the name of the last user action from the respective "interaction" event into the "actions.last" user model entry.

The architecture supports more complex operations. The outcomes of computations may again serve as sources for computations of other entries. In Figure 5.7.C, user action prediction illustrates the use of complex data types and derivations. We present the implementation of an algorithm in the user model as an example. Before discussing its implementation, we briefly summarize the Markov chain-based algorithm discussed in Section 3.4. Equation 5.1 shows the algorithm for computing a prediction of the next user action.

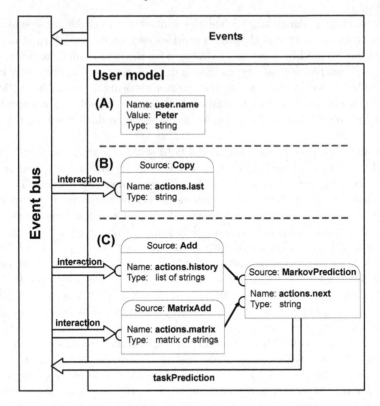

**Fig. 5.7.** An example user model with derivations.

$$s(t) = \sum_{j=1}^{N} d(j)h(t-j)A^j \tag{5.1}$$

$s(t)$ represents a probability vector of the next user action. The interaction history of the user is stored in a vector $h$ of length $N$. $d$ is a dampening factor that assigns more weight to more recent actions. A matrix $A$ stores the transition probabilities between actions and the matrix is computed from another matrix $T$, which stores transition counts between actions. This algorithm is implemented as follows in the framework (see Figure 5.7.C). The user model stores the list of past interactions a user model entry "actions.history" and updates this entry with the `Add` source, which appends interactions to the list. Moreover, the matrix "actions.matrix", which corresponds to matrix $T$, stores the number of transitions between two interactions and the `MatrixAdd` operator updates the matrix when a new "interaction" event is observed. The `MarkovPrediction` source of the "actions.next" entry uses these two entries to compute the most likely next interaction with the prediction algorithm. This operator converts the matrix "actions.matrix" into a probability matrix and combines it with the interaction history stored in "actions.history" to compute the prediction.

In this fashion, arbitrary derivations, such as modeling the user's favorite channel from the respective event parameters or predicting the most likely entry in a list, can be performed with values extracted from events by combining (and possibly creating new) operators. Other algorithms, such as the algorithms presented by Zukerman and Albrecht (2001), can be implemented similarly within this framework for deriving information directly from observed user behavior. In addition, domain-specific algorithms may be implemented. An extension mechanism allows the definition of custom operators as well as specific data types.

In addition to regular entries, the user model supports a special type of entry called property. Properties store user-dependent values for different components of the interactive system based on IDs of the interface elements. For instance, a property "isEnabled" may store for every element of the interactive system, such as states, graphical elements, or speech output prompts, whether they are enabled.

### 5.3.3 Connecting the User Model and the Ontology

Ontologies are a useful technology for user modeling (Razmerita et al., 2003; Heckmann et al., 2005). Semantic user models use inference techniques and may incorporate external ontologies. Since this framework comprises both a semantic layer and a user modeling component and the semantic layer contains information about the user model, using semantic user modeling techniques in this framework would be obvious. As discussed before, the user model requires however an efficient representation of data, such as matrices, which a semantic representation does not fulfill. Instead, this framework connects the user model and the ontology by means of bridging mechanisms.

User model entries correspond to properties of individuals. For instance, the user model entry user.name introduced in Figure 5.7.A corresponds to the property hasName of the individual CurrentUser, which is an instance of the User class. Therefore, the bridging mechanism synchronizes data stored in the user model and the corresponding statements in the semantic layer. This framework provides two methods for connecting the user model to the semantic layer. First, the user model entry definition comprises mapping information, which defines the ontology class and property of the corresponding individual, and a query is used to retrieve the respective individual from the ontology. A bridging component synchronizes the user model and the semantic layer based on this mapping information. Only user model entries with mapping information are synchronized with the ontology. Second, a special functor umProperty allows access to user model data from queries. In doing so, the query may select individuals from the semantic layer based on values from the user model. For example, a query may include this functor to load the name of the user into the query. These methods connect the user model and the ontology and thus facilitate a semantic notation and efficient data storage at the same time. Since both the user model and the OWL formalism exploit the XML Schema datatypes (XSD) (Biron and Malhotra, 2004), no data conversion between the respective types is required.

## 5.4 Applying Adaptation Patterns

In this section, we present an adaptation component that exploits both the semantic layer and the user modeling component to perform adaptations of the interactive system using the adaptations presented in Chapter 4. Rules have been used as a method for implementing adaptations (cf. Chu-Carroll, 2000 and Stephanidis et al., 1998). This framework also supports rules that connect specific events emitted by the user modeling component to program code, such as code defined in a scripting language. For instance, the observation of a specific user action in a specific screen may change the user's proficiency from beginner to expert. In doing so, the developer implements adaptations by constructing a rule base. However, a rule-based approach is not flexible and larger sets of rules become confusing. In addition, it does not support reuse of adaptation code well, because rules do not create an abstraction. Instead, we devised a more flexible approach that allows both a reuse of the adaptation definitions and an adjustment to a specific system.

We present an adaptation component that exploits the semantic representation of the interactive system to apply adaptations to the interactive system. A user modeling component, which we introduced in the previous section, observes the user-system interaction and employs different algorithms to model user behavior. The results of the user modeling procedure are forwarded to the adaptation component as adaptations triggers. These include a prediction of the next user action or an update of a user model entry that describes a favorite value, such as a TV channel or the cell phone number. After such a trigger was sent, the adaptation component selects an appropriate adaptation and an interface element that should be adapted. In the remainder of this section, we discuss the representation and execution of adaptations in the adaptation framework.

### 5.4.1 Representation of Adaptations in the Framework

Adaptations are represented by a set of abstract adaptation selectors and system-specific adaptation executors. System-independent selectors define adaptations on an abstract level. They connect a user model trigger, the selection of an individual from the ontology, and the selection of an abstract adaptation. Second, system-specific executors define the execution of an abstract adaptation on a specific interface element. Adaptation selectors allow a system-independent definition of adaptations, which may be used in different interactive systems. The system-specific adaptation executors enable a tight integration of the adaptation into different interactive systems. In the following, we discuss selectors and executors in more detail.

System-independent *adaptation selectors* define adaptations on an abstract level. They may be reused between different interactive systems that employ this framework. Selectors do not contain information on the execution of adaptations. They are connected to the user modeling component through an event trigger, for instance the prediction of an action by the user modeling component. When the respective event occurs, the adaptation component activates the adaptation selector and a query selects individuals from the ontology as adaptation candidates. For instance, one adaptation selector might select a graphical button that is connected to the predicted user action. Finally, the adaptation selector has an abstract adaptation associated to it that is executed on the selected individuals. For instance, the graphical button that triggers the predicted action might be emphasized by means of the "Component Emphasis" adaptation. Thus, an adaptation selector consists of an event trigger, a knowledge base query, and a reference to an adaptation. Since all this information does not depend on a specific interactive system, adaptation selectors may be reused between different interactive systems.

*Adaptation executors* define the execution of an adaptation on a specific interface element. Each executor defines the effects of an abstract adaptation on a specific type of interface element. For instance, the previously mentioned "Component Emphasis" adaptation is executed for a graphical button in interactive system "A" by updating the "width" and "height" properties to increase the size of the button. On interactive system "B", the adaptation of the button widget might be implemented as a change of the background image. Therefore, whereas executors may be reused between different systems, specific executors allow an integration that takes into account system-specific design guidelines.

Each of the adaptation patterns in Chapter 4 may be represented by a set of adaptation selectors and executors. More than one adaptation selector defines a specific adaptation, for instance for different user modeling events. In addition, a set of adaptation executors defines the adaptation for different interactive systems. We defined a set of adaptation selectors as well as default adaptation executors for each adaptation pattern. In the following section, we discuss how the adaptation framework carries out adaptations based on the specification introduced in this section.

Figure 5.8 illustrates the definition of adaptations in this framework. Adaptation selector "Selector A" is triggered by the prediction of an action and selects graphical buttons that trigger the predicted action. The selector recommends the abstract "ButtonEmphasis" adaptation for the selected buttons. The adaptation component has two executors for the "ButtonEmphasis" adaptation at its disposal. Executor "Executor a1" visualizes the adaptation by increasing the size of the button, whereas executor "Executor a2" changes the background color of the button. Adaptation selectors "Selector B1" and "Selector B2" demonstrate that different selectors may trigger one executor.

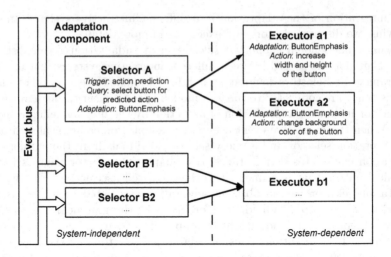

**Fig. 5.8.** The definition of adaptations is separated into system-independent adaptation selectors and system-dependent adaptation executors.

### 5.4.2 Executing Adaptations

The adaptation framework provides two methods for executing adaptations based on the selectors and executors: specification by the developer and a fully automatic execution by an adaptation component. While both methods follow a similar approach, the adaptation component derives the decisions taken by the developer automatically from the description of the adaptations. The execution of an adaptation consists of the selection of a combination of an adaptation and an interface element. For this purpose, a user modeling component triggers an adaptation selector.

In the first execution method, adaptations are assigned to interface elements by the developer. For this purpose, the developer assigns adaptation selectors to specific interface elements, for instance a "Component Emphasis" selector for a specific graphical button. Based on the information from the semantic representation, the development environment provides tool support by only showing matching adaptations for selected interface elements. For instance, if the designer selects a graphical button, only adaptations that are applicable to graphical buttons are presented. In addition, the system designer may influence the selection of adaptation executors by choosing a set of possible executors. However, the developer may define new selectors if no appropriate ones exist. In the second adaptation method, an adaptation component automatically selects a suitable combination of interface elements and adaptation selectors. For this purpose, the adaptation component uses the information encoded in the semantic representation to derive appropriate adaptation executors. These are in turn executed on the selected interface element. In doing so, the adaptation component automatically performs the decisions taken by the system designer.

The execution of the "Component Emphasis" adaptation pattern introduced in Section 5.4.1 serves as an example. Each button in a graphical interactive system executes a specific action. If the user modeling component predicts a user action, the "Component Emphasis" pattern highlights a button that triggers the respective action. For this purpose, an adaptation selector "Selector A" (see Figure 5.8) reacts to action predictions from the user modeling component by selecting all interface elements that execute the predicted action. The adaptation component decides whether an adaptation helps the current user and whether the selected adaptation is appropriate based on the selected interface element and context. Next, an adaptation executor is selected for each interface element. In the example, the adaptation component may decide between the executors "Executor a1" and "Executor a2". In this case, executor "Executor a2" is selected that highlights the graphical button by selecting a more noticeable background color. Thus, the adaptations may either be selected by the developer or derived automatically by an adaptation component.

## 5.5 Implementation

In this section, we introduce a reference implementation of the adaptation framework. The implementation shows the feasibility of the approach and serves as a test bed for an evaluation of the user modeling algorithms and adaptations. In addition to discussing the implementation, this section presents the steps necessary to apply the framework to a non-adaptive interactive system.

### 5.5.1 Reference Implementation

In order to investigate the feasibility of this framework as well as the presented adaptation approach, we developed a reference implementation of the adaptation framework. For this purpose, the framework was implemented as an extension of a modeling tool called EB GUIDE Studio (Goronzy et al., 2006). This tool is used to specify multimodal interactive systems. It comprises a simulation component for executing the interactive system, which serves as a dialog manager. The tool is implemented on the Java platform and provides an extension mechanism through plug-ins. EB GUIDE Studio employs a statechart formalism for the dialog logic and attaches both graphical and speech components to states.

The simulation component of the modeling tool serves as a dialog manager. In order to run the interactive system independently of the modeling tool on a target platform, a code generator usually exports the model of the interactive system to source code, which is in turn compiled into a target application.

Therefore, the target application also has to provide adaptation mechanisms. Since the adaptation framework has been implemented as a separate component, a target platform can integrate the framework and thus perform user modeling and adaptations independent of the modeling tool.

The adaptation framework extends the modeling tool with a semantic layer as discussed in Section 5.2. The layer first loads the OWL ontology and a set of ontology connectors instantiates the ontology classes to form the semantic layer. This layer is available both at design time and runtime of the system. The semantic layer is based on the Jena framework[6], a Semantic Web library with OWL support. In addition, the framework comprises a user modeling component. This component describes the user-system interaction and provides information about the user, such as detecting and predicting user actions and preferences of the user. All basic observations from the interactive system as converted into events and submitted to the event bus. The user modeling component includes an implementation of different algorithms, such as action detection and prediction algorithms. A user model applies further algorithms to this data. For this purpose, arbitrary algorithms may be implemented and custom data types and operators may be provided as Java extensions. Finally, the user modeling component provides adaptation triggers to an adaptation component. This component comprises adaptation executors and selectors for the adaptation patterns presented in this book. The prototype supports both adaptation methods from Section 5.4. In the following, we discuss the implementation of adaptations in this framework in more detail.

Different adaptive interactive systems have been developed with this framework. They implement adaptations such as providing adaptive help or emphasizing interface elements based on a prediction of the next user-system interaction. For this purpose, reusable adaptation selectors and system-specific adaptation executors have been defined for the individual adaptation patterns. In the following, we present the implementation of an adaptation selector and an adaptation executor for the "Component Emphasis" adaptation as an example. An XML notation is used for the definition of selectors and executors.

Figure 5.9 presents the definition of an adaptation selector called "ButtonEmphasisSelector" for the "Component Emphasis" pattern. The selector chooses a graphical button that triggers an action. The user modeling component predicted this action. First, a range defines the types of interface elements that the adaptation selector supports, in this case a button ("ds:Button" in terms of the ontology, lines 3–5). Next, an adaptation trigger connects the adaptation selector to the user modeling component by means of a trigger event (lines 6–8). When the trigger event is observed, the selector is activated. In this case, the adaptation trigger is activated if an "interactionPrediction" event is observed. In order to allow a flexible selection of events, event definitions may use wildcards. A number of adaptation candidates are selected from the knowledge base by a query, which uses a syntax similar to the SPARQL

---

[6] The Jena Framework: http://jena.sourceforge.net

```
01 <adaptationSelector name="ButtonEmphasisSelector"
02                pattern="ComponentEmphasis">
03   <ranges>
04     <range name="ds:Button" />
05   </ranges>
06   <trigger>
07     <event type="interactionPrediction" name="*" />
08   </trigger>
09   <subject>
10     <query variables="?element">
11 (?element rdf:type ds:Button)
12 (?triggerSource ds:hasName ?name)
13 (?element ds:triggers ?triggerSource)
14     </query>
15   </subject>
16   <adaptation name="ButtonEmphasis" />
17 </adaptationSelector>
```

**Fig. 5.9.** Example of an adaptation selector for the "Component Emphasis" adaptation pattern.

language. The query includes values from event triggers, in this case the name of the prediction in the "?name" variable. For this purpose, the value of the event parameter is assigned to the respective variable. The query in the example selects all interface elements that are buttons and trigger the predicted action (lines 9–15). Finally, the selector includes the name of an abstract adaptation. Different executors may be defined for this abstract adaptation. In this case, the abstract adaptation is called "ButtonEmphasis" (line 16).

Figure 5.10 presents an adaptation executor that complements the previously presented adaptation selector. The executor defines an adaptation ("ButtonEmphasis") for a specific type of interface element, in this case generic interface components ("ds:Widget" as the generic type for all graphical interface elements, line 3). The executor may be restricted to a specific kind of context, such as a graphical ("view") or speech ("speech") context. The core of the adaptation executor is an executor element, which the system designer may extend. In this case, a type called "ActionExecutor" changes properties of interface elements. The presented executor increases position and size of the interface element (lines 4–17). Similarly, the adaptation executor could implement other visualizations, such as a change of the background color or an animation.

However, the adaptation component does not apply the adaptations directly to the interactive system when the respective adaptation selector is activated. The selector may be executed while the respective interface element is not activated. For instance, a computation in the user modeling component may cause a delay. Therefore, the context of an adaptation executor serves as a condition. It may be of the type "view" or "speech". A context condition

```
01 <adaptationExecutor name="SimpleButtonEmphasis"
02          adaptation="ButtonEmphasis" context="view">
03   <range class="ds:Widget" />
04   <executor class="ActionExecutor">
05     <widgetproperty name="x">
06       <translate value="-3^^int" />
07     </widgetproperty>
08     <widgetproperty name="y">
09       <translate value="-3^^int" />
10     </widgetproperty>
11     <widgetproperty name="width">
12       <translate value="6^^int" />
13     </widgetproperty>
14     <widgetproperty name="height">
15       <translate value="6^^int" />
16     </widgetproperty>
17   </executor>
18 </adaptationExecutor>
```

**Fig. 5.10.** Example of an adaptation executor for the "Button Emphasis" adaptation. An executor complements the adaptation selector from Figure 5.9.

is valid if the current graphical screen or speech component corresponds to the adaptation element. For instance, if a graphical button should be adapted and the adaptation context is "view", the adaptation is only executed if the view of the button is active.

Adaptations are not always applied when they are selected, but only when the respective context is active. Figure 5.11 shows the adaptation procedure. If an adaptation is triggered and the context is valid, the adaptation is executed. Otherwise, the adaptation is added to a list of active adaptations and executed when the context becomes active. If the context becomes inactive, the adaptation is reverted and added to the list of active adaptations. This behavior allows an adaptation to be executed whenever a context is entered. For example, this approach allows a help message to be displayed whenever the user enters a specific screen. The selection of the adaptation is decoupled from its application.

### 5.5.2 Applying the Framework to a Non-Adaptive Interactive System

In the following, we discuss the steps necessary to apply the adaptation framework to a non-adaptive interactive system. In doing so, we further illustrate the workings of the framework. As a prerequisite, the system has to be developed in an environment supported by the adaptation framework, such as the reference implementation presented in this section. The system description in the semantic layer is derived automatically from the model-based description.

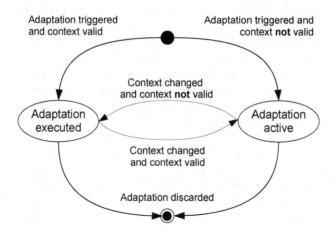

**Fig. 5.11.** The adaptation component has available a list of possible adaptations and executes them when the respective contexts become valid.

The configuration of the user modeling component comprises the following steps: First, the interactive system has to emit events to the event bus, for instance caused by a remote control or speech input. In order to describe higher-level user behavior, the developer has to provide a provision for recognizing user actions, for instance by training PDFA automata, and a configuration for the task model. Finally, the developer configures the user modeling component such that it employs algorithms to model user behavior. This includes for example predictions about user actions and preferences. For this purpose, the developer uses existing algorithms or adds new ones.

In order to support adaptations, interface elements first have to be prepared such that the adaptation component may update them. For this purpose, interface elements have to read a per user configuration either from the user model or a knowledge base. In addition, they may offer properties that the adaptation component changes. For instance, a list that highlights the user's favorite items may read an entry from the user model that encodes the user's preferences. The user modeling component may alter a property of the list to encode the user's favorite items. The selected approach depends on the specifics of an interactive system. In addition, the developer has to provide information about the types of the interface elements, such as list or button. Next, if no matching selector is available, the adaptation selector has to be defined. However, selectors can be reused between different interactive systems. Finally, custom adaptation executors have to be provided. They consider the specific properties of the individual interface elements, such as how user modeling information is communicated to the elements. The framework includes a set of default executors, which only modify properties that are common to all interface elements (e.g. position information).

In the remainder of this section, we present two use cases that illustrate the application of adaptations to an interactive system. In the first example, speech output prompts shall be adapted to the user's expertise by means of the "Alternative Elements" adaptation. First, the system designer creates a group of speech output elements, for instance a long welcome text, a short text, and a beep, including an annotation that assigns the prompts to beginners, advanced users, and experts respectively. The framework comprises an adaptation selector that is activated when the interactive system opens a different speech context. An adaptation executor updates user model properties to enable or disable the appropriate prompts. The interactive system consults the user model to play back the prompts selected by the adaptation.

A second example discusses the adaptation of an interactive system by emphasizing a graphical button to guide the user to the next interaction step. The emphasis highlights the button by selecting a more noticeable background color and increasing the size of the button. First, the developer adds annotation to each interface element, such as buttons, to specify which action it triggers. The framework comprises an adaptation selector that chooses interface elements as adaptation candidates based on action predictions by the user modeling component. Finally, the developer creates a custom adaptation executor for the requirements of this specific interactive system. In this case, the executor updates properties of the button to increase the size and change the background color. Alternatively, the adaptation selector may use an animation to display the emphasis. Thus, the adaptation component provides for a flexible specification and execution of adaptations.

## 5.6 Discussion

In this chapter, we presented a novel adaptation framework that provides a reusable shell for the user modeling algorithms and the adaptation patterns introduced in previous chapters. The aim of the framework is to support the development of adaptive interactive systems. This framework comprises several components, which were discussed in this chapter. First, a semantic layer constitutes the foundation and creates an abstract representation of the interactive system as well as other building blocks, such as the user and the user-system interaction. This layer is based on semantic web technologies, such as OWL. Second, a user modeling component collects basic observations from the user-system interaction and distributes them among different components, such as an interaction model and a task model. A user model computes derivations from these events, such as a prediction of the next user action or a model of the user's preferences. An implementation of the user modeling algorithms discussed in Chapter 3 illustrates the use of the user model. Third, an adaptation component employs a formalization of the adaptation patterns from Chapter 4 to perform adaptations to the interactive system. The description of adaptations consists of adaptation selectors and adaptation executors.

The framework may be reused in different interactive systems. We presented a reference implementation of the framework that proves its feasibility. In addition, we illustrated the workings of the framework with examples.

In order to prove the feasibility of the user modeling algorithms and adaptation patterns as well as the adaptation framework, we present a user evaluation in the following chapter. Different interactive systems use this framework to implement adaptations. Thus, the framework presents a viable and reusable foundation for the implementation of user-adaptive interactive systems.

# 6

# Evaluation

The only man who behaves sensibly is my tailor;
he takes my measurements anew every time he sees me,
while all the rest go on with their old measurements
and expect me to fit them.

—George Bernard Shaw (1856–1950)

Research on adaptive interactive systems has produced numerous adaptive systems, such as hypertext, office, or speech dialog systems (see Chapter 2). However, little empirical evidence exists on topics such as the effect of adaptations on the user-system interaction and the decision when and how to apply adaptations. For example, Lavie et al. (2005) criticize this lack of empirical foundation. In order to allow developer to implement adaptations properly, evaluation evidence has to be provided that is applicable to all interactive systems. In this chapter, we present the results of user tests for user modeling algorithms and adaptations. The evaluations have been performed such that the results may be transferred to different interactive systems. In doing so, we contribute to building a body of generic adaptation evidence.

Research on the evaluation of adaptive interactive systems revealed that an assessment of the individual components provides more insight than an evaluation of the complete system. Otherwise, the developer is unable to differentiate between good and weak components. In this work, we evaluated the individual components separately, namely user modeling algorithms and adaptations. Moreover, an evaluation of the individual components enables the reuse of facts that have been discovered in an evaluation. For example, an interactive system may select user modeling algorithms and adaptations that have been evaluated independently of each other. In this chapter, we present an evaluation of the individual components of adaptive interactive systems that we introduced in previous chapters. We investigated the attitudes of users towards adaptive interactive systems. In addition, we performed an evaluation of different user modeling algorithms. Finally, we implemented a set of interactive systems that provide different adaptive features. The evaluations have been performed such that the results are generic. However, an assessment of the individual components does not replace an evaluation of an end-to-end system, which uses a specific combination of adaptations and user modeling algorithms.

For some parts of this chapter, data from the evaluation of a prototype of the DICIT project (Matassoni et al., 2008; Marquardt et al., 2008) was used. The DICIT project (Distant-talking Interfaces for Control of Interactive TV), an European FP6 project, developed a speech-enabled digital TV system and supports distant-talking natural language interaction in addition to a remote control. The prototype offers the general TV functionality, i.e., watching TV or changing channels and volume. In addition, an electronic program guide (EPG) allows users to browse the TV program and select shows by changing a set of filters, such as time or channel. Users may put interesting programs on a watch list and change settings, such as switching on or off the speech output. All functions are controlled either by remote control or speech input. The DICIT system provides a graphical interface and plays speech output prompts to communicate with the user.

The chapter is structured as follows. First, an introduction to evaluation approaches for adaptive interactive systems is given. Next, attitudes of users of adaptive interactive systems and a larger group of international users are investigated. Thereafter, an evaluation of the user modeling algorithms from Chapter 3 is presented. Finally, an evaluation with test subjects of interactive systems that implement the adaptations from Chapter 4 provides empirical evidence for the application of the adaptations.

## 6.1 Evaluating Adaptive Interactive Systems

In this section, we provide an overview of different evaluation approaches for adaptive interactive systems. In doing so, we explain the reasoning behind evaluating the individual components separately. A general introduction to adaptive systems and the most important components is given in Section 1.2. In the remainder of this section, we discuss evaluation approaches for interactive systems that comply with these models.

The layered evaluation approach by Brusilovsky et al. (2001) proposes a separation of the evaluation of the interaction assessment and adaptation decision making processes. By investigating both components at the same time, a positive evaluation result of one component may obscure problems in another. Moreover, problems discovered by the evaluation do not reveal the source of the problems. If however the components are evaluated separately, problems in both components may be identified and addressed subsequently. The application to a tutoring system called InterBook illustrates the layered evaluation approach. Brusilovsky et al. conclude that having applied this approach in previous evaluations could have made problems visible earlier.

Paramythis et al. (2001) present a more fine-grained evaluation approach for adaptive systems. It is based on a decomposition of adaptive systems into different components. The evaluation is performed on different groups of these components. Breaking down a system into its constituents facilitates an assessment of the individual parts and enables a reuse of the evaluation results for different systems. A decomposition of adaptive systems into different components forms the basis of the approach by Paramythis et al. For instance, these include interaction monitoring, explicitly provided knowledge, the adaptation decision, and an automatic assessment of the adaptation. Related components form evaluation modules, which cover coherent topics. For example, module $A1$ comprises "interaction monitoring; interpretation and inferences; and user modeling" and therefore describes the user modeling process. Module $C$ on the other hand is "applying adaptations", i.e., the decision whether an adaptation should be performed. Modules A1 and C correspond to these parts of adaptive interactive systems which we evaluate in this chapter. However, the individual adaptations are not part of the evaluation modules in the decomposition approach. The most obvious evaluation approach is the "with-and-without" approach. It compares an adaptive interactive system to the same system without adaptation. Adaptation can however not be disabled if it is an integral part of the system. Moreover, the with-and-without evaluation compares two systems that possibly are not optimal and thus produces distorted results. Since the adaptations presented in Chapter 4 are not necessarily an integral part of the system, a "with-and-without" evaluation still allows valuable insight into the use of these adaptations.

The evaluation approach presented by Paramythis and Weibelzahl (2005) builds on the module-based approach by Paramythis et al. (2001), but presents a process-oriented decomposition of adaptive interactive systems. The adaptation process is separated into the following steps: a) collecting input data, b) interpreting the collected data, c) modeling the current state of the world, d) deciding which adaptation to apply, and e) applying the adaptation. A number of dynamic and static models, such as user model and system model, complement these stages. The individual steps and models are described in more detail in Section 1.2. Paramythis and Weibelzahl argue that the process-oriented decomposition poses a helpful partitioning for the evaluation of adaptive interactive systems. Although their approach is in early stages and has not been transformed into a full evaluation model, a comprehensive and comprehensible decomposition is still presented. The adaptation approach presented in this book employs a similar model.

## 6.2 Attitudes of Users towards Adaptive Interactive Systems

In addition to objective measures, the attitudes of users play an important role with regard to user acceptance. For this purpose, we performed two user surveys of adaptive features in interactive systems. The first one asked users of different adaptive systems. The second one questioned a larger group of users from different countries, who however had not used an adaptive system.

### 6.2.1 Users of Adaptive Systems

In order to investigate the adaptation patterns introduced in Section 4.3, we carried out a user test of different interactive systems that implement these adaptations. For this purpose, users had to perform a number of tasks with an adaptive and a non-adaptive version of different test systems. At the end of the evaluation, users were handed a questionnaire to collect their attitudes towards adaptive interactive systems in general. The questionnaire is given in Appendix B (Questionnaire G). In total, 36 users filled in the questionnaire; 20 of them had tested four different systems and another group of 16 had tested only a single adaptive interactive system.

The users answered each question by means of a 7-point Likert scale, with "1" corresponding to "agreement" and "7" to disagreement. The results of the questionnaire are shown in Figure 6.1. Users agree strongly with the statement that adaptivity is an interesting concept (Figure 6.1.b, median value of 1). The subjects expected that adaptivity would be used commercially in the future (Figure 6.1.b, median value of 1). In addition, users would not select a non-adaptive version of the system if they could select (Figure 6.1.c, median value of 6). Finally, subjects were undecided about the idea to relinquish other features in favor of adaptivity (Figure 6.1.d, median value of 4). Therefore, users of different systems showed a positive attitude towards adaptation in general. However, they would not happily relinquish other features in favor of adaptation.

### 6.2.2 Cultural Influence

The (Matassoni et al., 2008) developed a speech-enabled digital TV system, which includes an EPG. Users from different countries, namely Italy, Germany, the Czech Republic, and the United States of America, tested the final prototype of the DICIT system in an evaluation. The evaluation included a

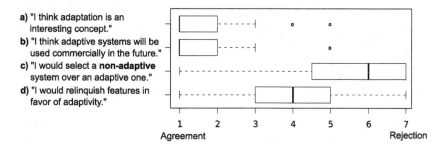

**Fig. 6.1.** Attitudes of users of adaptive interactive systems towards adaptivity. The users stated their general attitudes by means of four questions (a–d).

questionnaire that we extended for this work with a number of questions about adaptive features. Whereas the implementation of these features was not part of the DICIT project, the results still give an insight into the attitudes of people in different countries towards adaptivity. The full evaluation of the DICIT system is available as a public deliverable of the project (Sowa and Arisio, 2009). In total, 159 subjects participated in the user study, including 49 natives of Germany (group "DE"), 50 natives of Italy (group "IT"), and 23 natives of either the United Kingdom or the United States of America (group "EN"). The remaining subjects were non-native speakers from different countries and are excluded in this survey. The questionnaire used a 10-point Likert scale from "1" for a very negative answer to "10" for a very positive one.

The results of the questions from the DICIT questionnaire that address adaptivity are given in Figure 6.2. Significant differences between the three groups could not be observed. Question 21 asked about the ability of a system to adapt to user behavior in general. The users approved of such a feature with a median value of nine for all groups (Figure 6.2.a). Question 22 suggests a specific example of adaptation, namely highlighting the most frequently used functions on the screen automatically. The participants approved of this idea, but the attitude of German subjects (median value of 9) was more positive than the attitude of the subjects in other countries (Figure 6.2.b, median value of 8). Finally, question 23 asked whether a user modeling component that observes the user would make the subjects feel monitored. However, users do not feel uncomfortable because of the observation (Figure 6.2.c, median value of 9).

In summary, users in all countries expressed a positive attitude towards adaptive features in interactive systems and noted that an observation by a user-modeling component would not make them feel uncomfortable. No significant differences could be found between German, Italian, and English or American users. Thus, our findings from the two surveys presented in this section indicate an open-mindedness and positive attitude of users towards the concept of adaptivity in general. Whereas privacy is regarded as an important

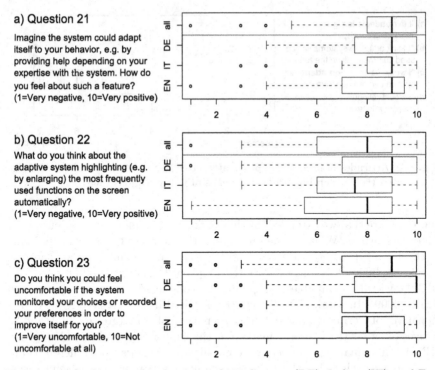

**Fig. 6.2.** Attitudes towards adaptivity of 122 German (DE), Italian (IT), and English or American (EN) users of the DICIT system, a speech-enabled interactive TV system.

topic in literature (cf. Kobsa, 2007), users did not express such concerns in this survey.

## 6.3 Evaluation of the User Modeling Algorithms

In this section, we present an evaluation of the user modeling approach presented in Chapter 3. This includes an algorithm for recognizing user actions by means of probabilistic automata and an approach for user action prediction by means of a task model or Markov chains. For this purpose, applied to algorithms to log data from user tests.

### 6.3.1 Test Data

We conducted the investigation of the user modeling algorithms presented in Chapter 3 with log data obtained during the evaluation of the DICIT project. This evaluation was conducted with 20 test subjects. The users had to perform a number of tasks, which cover all parts of the system. For this purpose, they used both remote control and speech interaction. The average length of the evaluation sessions is 24 minutes. During the evaluation, extensive log data was collected in log files. A log file is a sequence of interaction events, such as a key press, a speech utterance, or a reaction to user input by an interactive system. A short excerpt of a log file is given in Figure 6.3, in which a user presses a button (line 1) that causes different reactions of the system (lines 2–4). We annotated all log files with information about user actions using a custom annotation tool (see Wesseling et al., 2008). A screenshot of the tool is given in Figure 6.4. This tool displays recorded log files in timeline views and presents different event types separately. For instance, one timeline may show remote control events, whereas another one may display speech input. In doing so, the interaction may be investigated in detail. In order to add annotations, log event sequences that correspond to specific user actions are selected and marked with the name of the user action. While the assessment was performed within the digital TV domain, the algorithms are applicable to any kind of interaction log data.

### 6.3.2 Action Recognition

In Section 3.2, we introduced an algorithm for recognizing user actions that employs probabilistic automata. We tested this algorithm using the DICIT recordings by comparing a manual annotation to an annotation created by means of the PDFA models. One important metric of this algorithm is the amount of training data needed for a sufficiently well performing model. Therefore, the evaluation investigates the rate of correct matches in relation to the number of sessions used for training. For this purpose, PDFA models were trained with an increasing number of sessions (between one and 15) and tested with five sessions. The evaluation was performed by means of cross-validation with a random selection of sessions with 10 repetitions. In these recordings, 32 different action classes were identified, such as "show results" or "change channel". The sequence length of these action classes ranges from one to 64 events, with the average length being 3.1 events. A session consists of 141 interactions on average.

```
1 [1180520776220] hw name={COMM_0x6e}
2 [1180520776220] event name={Start}
3 [1180520776220] state name={Session_Main}
4 [1180520776376] view name={WelcomeView}
```

**Fig. 6.3.** Exemplary log lines, showing a button press (line 1) and the reactions of the interactive system to this input (lines 2–4).

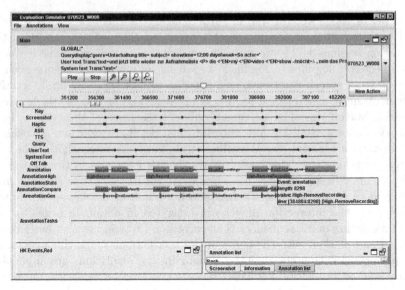

**Fig. 6.4.** An evaluation and annotation tool was used to perform the annotations of the log data.

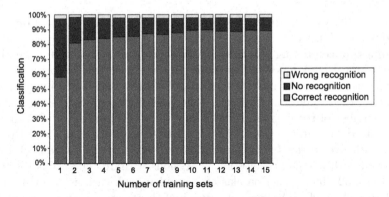

**Fig. 6.5.** Evaluation results of the PDFA matching algorithm. An increasing number of log sessions were used to train the matchers and five sessions were used for testing.

Figure 6.5 presents the results of the comparison of the automatic and the manual annotations in relation to the number of training sessions. Starting with a match rate of 58.1 % with one training session, the match rate increases with a growing number of sessions until it reaches a maximum of 89.5 % with 11 sessions. The match rate with two training sessions already amounts to nearly 80 %. Therefore, the presented approach works with a comparatively small number of training sessions and the match rate does not improve considerably after 10 sessions.

Matching fails for several reasons. First, the matcher does not trigger at all (*no recognition*). Since the action is not present in the training data, it appears in the manual annotation, but not in the automatic annotation. Using more training data therefore reduces these kinds of errors, as can be seen in Figure 6.5. Second, a wrong class is selected (*wrong recognition*), which occurs in only about 2 % of the evaluation data. This kind of error occurs because of inconsistencies or ambiguities in the annotations.

Therefore, PDFAs present a viable means for detecting user actions in a sequence of basic events. However, this approach has limitations. First, it highly depends on the annotations like all statistical approaches, i.e., only patterns occurring in the annotated sessions can be recognized. In addition, the consistency of the annotations affects the recognition accuracy. Second, actions that depend on the number and order of events are not reflected well by the statistical nature of this approach. For instance, the number of up and down key presses decides which element in a list is selected, but the probabilistic nature of the automata does not consider this well. In order to comply with the PDFA approach, selections of list elements need to create specific "list selection" events.

### 6.3.3 Action Prediction

In Section 3.4, we presented two approaches for predicting user actions, namely prediction based on a task model and Markov chains. We compare these two algorithms in the following. This evaluation exploits the data from the DICIT evaluation. Ten sessions were used for training and ten for testing. Since the behavior of real users is erratic, the prediction accuracy always remains significantly below 100 %. Hartmann and Schreiber (2007) report in a comparison of different prediction algorithms that the accuracy was limited to 40–60 %. The task model prediction requires that a task is currently active. Therefore, we determined the task model coverage, which describes the ratio of actions in a session for which a task was active. With a coverage of 99.3 %, a valid task was active during virtually all interactions in these recordings.

We conducted the evaluation as follows. The annotated interaction sessions were iterated sequentially and the task model tracked the action sequence. For each interaction, both algorithms performed a prediction, which we compared to the action the user actually performed. Based on these predictions, a match rate was computed. The Markov chain algorithm uses the task model information to eliminate invalid predictions. The rate of Markov predictions that were invalid but corrected was 1.3 %. As can be seen in Figure 6.6, the Markov prediction (59.8 % correct) was more accurate than the approach that uses only the statistical task model information (56.3 % correct). The computation time for the Markov prediction amounted to 8.3 ms and 0.2 ms for the task model prediction with a dated 1.6 GHz PC. On a more 3.0 GHz hardware , the average time was less than 1 ms for both algorithms. Whereas the difference between the two algorithms is not significant, the Markov prediction did not

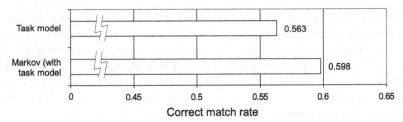

**Fig. 6.6.** A comparison of two action prediction algorithms, namely task model prediction and Markov prediction.

only produce better results, but also more variant predictions by not always recommending the same action for a certain state in the task model. Therefore, the Markov prediction achieves a better prediction rate. The prediction produces results that are more valuable for the user. Since the performance of both algorithms is comparable, the selection of the prediction algorithm may depend on other criteria, such as the available computing resources.

## 6.4 Evaluation of the Adaptation Approach

In order to investigate our adaptation approach, we conducted a user test for the adaptations presented in Chapter 4. These user tests examine the effectiveness of the adaptations and provide empirical evidence for the application of these adaptations. For this purpose, we created different test systems that implement these adaptations. Although each system was created specifically for the evaluation, they were derived from actual systems from two different domains: a digital TV with an EPG and an automotive dashboard interface. In this chapter, we present the results of a user test with these systems.

Design patterns receive validity when they are used repeatedly and therefore are considered efficient solutions. However, Metzker et al. (2003) argue that HCI knowledge, which is for instance captured by means of patterns, needs empirical validation. This chapter contributes to this endeavor by providing an empirical evaluation of systems that employ the adaptation patterns presented in Chapter 4. The evaluations not only investigate whether an adaptation improves the user-system interaction, but also give insight into the general conditions under which the adaptations perform well.

The evaluation of the adaptations was conducted with two groups of users and five different interactive systems. In a first evaluation, 20 users tested four different adaptive systems. In a second evaluation, 16 different users worked with one adaptive system. The evaluations comprise both a subjective survey by means of a questionnaire and an objective investigation by means of log data collected during the tests. The subjective measures include perceived attractiveness and usability. In addition to evaluating the different adaptations, the feasibility of the adaptation framework was demonstrated by implementing the adaptive test systems with the framework. For this purpose, the system uses the semantic layer and the adaptation component. The user modeling component was configured to predict actions and values according to the evaluation task. In the following, we introduce the test setup. Thereafter, we discuss the individual test systems and their evaluation.

### 6.4.1 Test Setup

A separate interactive system was created for the investigation of each adaptation pattern. The effects of the individual adaptations could not have been investigated as clearly with actual applications or multiple adaptations implemented in one system. All systems were implemented using EB GUIDE Studio (Goronzy et al., 2006), a modeling tool for interactive systems. The adaptation framework was employed to show its feasibility. Only the "Shortcut Area" adaptation did not use the framework. The adaptations were implemented by means of adaptation selectors and adaptation executors, as described in Chapter 5. In order to simulate the interaction with a remote control for a TV system or a push rotary device used in automotive dashboard systems, the users clicked on virtual buttons on the screen. In doing so, the number of basic interactions could be surveyed more precisely than with a touch-enabled interface, e.g. with regard to scrolling. All events caused by the user-system interaction were written to a log file to enable an assessment of the user's performance. All subjects used both an adaptive and a non-adaptive version of each interactive system to facilitate a comparison of the two conditions. The order in which the subjects tested the individual systems was rotated to eliminate an effect of the order, for instance through a training effect. Likewise, the adaptive and the non-adaptive version were used first alternately.

The test users completed a questionnaire to facilitate an assessment of their attitudes. The full questionnaire is shown in Appendix B. The subjects filled in a questionnaire both after the adaptive and the non-adaptive version of each system. Each questionnaire started with a set of opposed attribute pairs (such as "good" and "bad" or "predictable" and "unpredictable"). These attributes are derived from the AttrakDiff questionnaire (Hassenzahl et al., 2003), which measures the "pragmatic and hedonic quality" of a product. Each attribute pair was rated on a 7-point Likert scale. The attribute pairs were used to extract perceived attractiveness and usability measures. In addition, the questionnaire for the adaptive versions comprised a group of questions regarding the users' attitudes towards the adaptive features by means of three hypotheses: 1) the adaptation confuses the user (H1), 2) the adaptation accelerates the interaction (H2), and 3) the adaptation supports the user (H3). A total number of eight questions express the hypotheses. In addition, the questionnaire asked users whether they perceived the system as intrusive. Finally, users selected their preferred version (adaptive or non-adaptive) after having tested both versions. The evaluation was split into two parts: Four

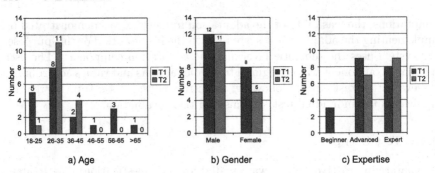

**Fig. 6.7.** The demographical characteristics of the two test groups T1 and T2 that were used for the evaluation of the adaptations.

adaptations ("Component Emphasis", "Adaptive Help Presentation", "List Element Selection", and "Alternative Elements") were tested in user test T1 and the "Shortcut Area" adaptation was tested in a separate user test T2. In total, 20 German subjects participated in evaluation T1 and the language of the respective interactive systems was German. Twelve subjects were male, eight female. The age, gender, and expertise distributions are given in Figure 6.7. With regard to age, more than half of the subjects were 35 years old or younger. In a self-assessment, three subjects regarded themselves as beginners, nine as advanced users, and eight as experts. For each test system, the users had to perform tasks that were explained on a worksheet. In addition, the users received a printed documentation of the respective system and a short oral introduction by the supervisor. The tasks were similar for the adaptive and the non-adaptive version to facilitate a comparison between both versions. Rather than performing actual user modeling, staged predictions that correspond to the respective tasks were used to assess the adaptations. Therefore, all user modeling predictions perfectly predicted the next user action. While this accuracy cannot be achieved in real-world user modeling systems, the effects of the adaptations become more obvious.

User test T2 investigated the "Shortcut Area" adaptation and was conducted prior to test T1 with different subjects. Since the adaptation framework was still incomplete at that time, the system did not employ the framework. However, the algorithms and adaptations have since then been implemented in the adaptation framework. Sixteen German test subjects participated in evaluation T2. The questionnaires were the same as in evaluation T1 (see Appendix B). As can be seen in Figure 6.7, all users in this group were younger than 46 years and 11 out of 16 users were between 26 and 35 years old. The number of male participants (11) was considerably higher than the number of female subjects (five). Moreover, this group consisted only of advanced users (seven) and experts (nine). In the following, we present the results of the user tests of the five different adaptive interactive systems.

**Fig. 6.8.** The main screen of the test system for the "Adaptive Help Presentation" adaptation. The help message reads: "Press red button to open list of shows."

### 6.4.2 Adaptation Pattern: Adaptive Help Presentation

The "Adaptive Help Presentation" adaptation supports the user by providing help that is tailored to the current user and situation. A digital TV system served as a test system for this adaptation.

**Test System and Task**

We tested the "Adaptive Help Presentation" adaptation using an interactive system that resembles a digital TV system with an EPG. A screenshot of the main screen is given in Figure 6.8. The user browses a list of TV shows, which is limited to 85 entries, and puts shows on a watch list. The TV system notifies the user if a show starts that is on the watch list. In addition, the user can change settings in a screen, such as the font size in the list of shows. The user controls this system by means of up and down buttons and three color buttons (red, green, and yellow). The meaning of the color buttons changes in every screen, with the respective functions being explained on the bottom of the screen.

The adaptive version of the test system displays a yellow box with a concise help message in the upper right corner of the screen. This message explained the user the next action according to the assignment sheet. The instruction in Figure 6.8 tells the user to press the red button to open the list of shows. The help messages relate to the current task. To improve the visibility of the help messages, the help box is faded in with a delay of two seconds.

The users performed different tasks in a given order: putting a show from the result list on the watch list (two times), removing an element from the scheduling list (two times), changing one setting (font size from regular to big), resetting the settings, and clearing the scheduling list. Similar tasks were selected for the adaptive and the non-adaptive version, but in a different order and with different values. However, the minimum number of interactions

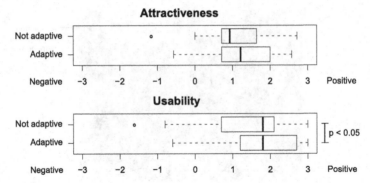

**Fig. 6.9.** Attractiveness and usability ratings of the "Adaptive Help Presentation" adaptation.

required to complete the tasks in both versions was constant in order to allow a comparison of the two variants. The printed documentation for this test system was the most extensive of all systems due to the complexity of the test system and the task.

### Subjective Evaluation

Figure 6.9 gives the views of the users with regard to attractiveness and usability. On average, the subjects rate the adaptive version as slightly more attractive than the non-adaptive version. The usability of the adaptive version is assessed as significantly better ($p < 0.05$).

With regard to the three hypotheses (see Figure 6.10), differences between expert users and non-expert users can be observed. Users perceive the adaptation as not confusing (H1, median value of 6.50), as faster (H2, median value of 2.00), and supportive (H3, median value of 1.25). However, the view of expert and non-expert users differ significantly. Whereas non-expert users regard the adaptation as not confusing, experts perceive the adaptation as neutral with regard to confusion (H1, median value of 7.00 vs. 4.63 with $p < 0.01$). Similarly, non-expert users opine that the adaptation accelerates the interaction, whereas experts see little influence of the adaptation on the performance (H2, median value of 1.25 vs. 4.00 with $p < 0.05$). Non-experts perceive the adaptation as supportive (H3, median value of 1.00 vs. 2.50 with $p < 0.05$), but expert users approve of this statement less. Therefore, non-expert users rate the adaptation significantly better than expert users. All users rejected the notion that the system was intrusive strongly with a median value of 7.00.

A majority of 13 out of 20 users prefer the adaptive version, five users prefer the non-adaptive version, and two users are undecided. However, the five users who prefer the non-adaptive version are all experts.

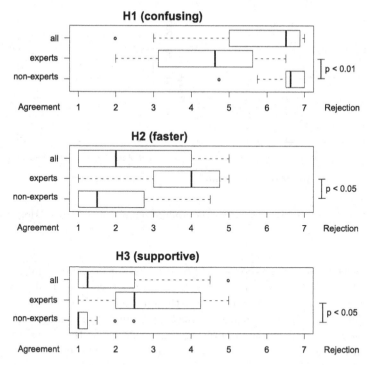

**Fig. 6.10.** User ratings for the three adaptation hypotheses for the "Adaptive Help Presentation" adaptation.

### Objective Evaluation

For the objective evaluation, the fastest user (V7) and the slowest user (V2) were removed from the set of log files. The adaptation significantly reduces the number of interactions ($p \leq 0.05$) compared to the non-adaptive version. This finding proves that users consult the adaptive help rather than using "trial and error", which they rely on in the non-adaptive version (Figure 6.11).

However, the operating time of the adaptive version remains constant compared to the non-adaptive version (Figure 6.11). We attribute this to a strong and significant ($p \leq 0.05$) learning effect between the first and the second version. Regardless of whether the adaptive or the non-adaptive version is used first, the task is solved more quickly with the second version. In addition, the reading time of the help message affects the operating time of the adaptive version negatively.

### Discussion and Lessons Learned

Beginners and advanced users regard the "Adaptive Help Presentation" adaptation as significantly more helpful and positive than expert users. Therefore,

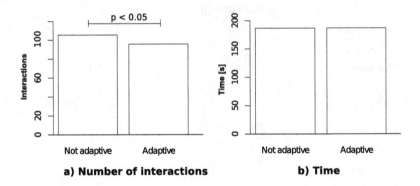

**Fig. 6.11.** Objective measures from the evaluation of the "Adaptive Help Presentation" adaptation.

the adaptive help feature should only be enabled for beginners and advanced users and disabled for experts. The adaptation improves the number of interactions significantly, but does not affect the interaction time. A more detailed study, which eliminates the learning effect, is required to investigate the influence of adaptive help on the operating time. Therefore, the "Adaptive Help Presentation" adaptation assists beginners and advanced users who work with interactive systems.

Whereas unintrusiveness was a design goal for the test system, a pre-test study revealed that users overlook help messages unless the help is presented with sufficient visibility. Color, location, and fading in turned out to be viable means for controlling the visibility and intrusiveness of the help box. In addition, users do not read help messages unless these are concise and clear. Finally, the complexity of the task has to warrant help. Otherwise, people either do not read the help messages or are even annoyed.

### 6.4.3 Adaptation Pattern: Component Emphasis

The "Component Emphasis" adaptation supports the user by drawing attention to interface elements through emphasis. We evaluated this adaptation using a digital TV system.

### Test System and Task

The test system is conceptually part of a digital TV system, which allows the user to browse an EPG. In order to cope with the vast number of shows, the user selects a number of search criteria, namely channel, time, date, and genre, to constrain the number of visible shows. A screenshot of the main screen of the test system is given in Figure 6.12. The user opens up a list of possible values for each category and selects a value, e.g. "Sat.1" in the list of channels

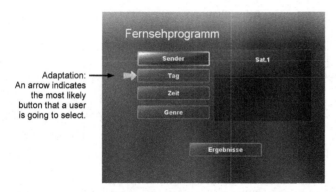

Adaptation:
An arrow indicates
the most likely
button that a user
is going to select.

**Fig. 6.12.** The main screen of the test system for the "Component Emphasis" adaptation. The user already has selected a value for the channel criterion.

("Sender"). After having specified a set of criteria, the user opens the result screen by activating the "Results" ("Ergebnisse") button. The result screen however is only a mockup. The user controls the test system by means of up, down, OK, and back buttons.

The adaptation inserts a yellow arrow on the left-hand side of the button that the user is supposed to select next according to the task. For each version, the user had to execute four tasks that each consist of selecting three different filter criteria and opening the result screen. For instance, one task instructed the user to select "RTL" for channel, "Friday" for day, and "documentary" for genre. The number of actions required for each task was kept constant by balancing the list positions of the individual filter values. In doing so, the number of interactions and the execution time could be compared between the different tasks.

### Subjective Evaluation

Only a small yet not significant improvement can be observed with regard to attractiveness and usability (Figure 6.13). However, the adaptive version is judged very positively (Figure 6.14): The idea of the system as being confusing (H1) is rejected with a median value of 6.75. Users strongly agree with a median value of 1.00 with the statement that the adaptation decreases the interaction time (H2). Moreover, users agree with a median value of 1.00 that the adaptation supports them (H3). Users reject the statement that the adaptation is intrusive with a median value of 6.00. 70% of the users (14 out of 20) prefer the adaptive version, 20% (four users) prefer the non-adaptive version, and 10% (two users) are undecided.

### Objective Evaluation

Users had to perform four subtasks for the adaptive and the non-adaptive version. For the objective assessment, we removed the first task, because it

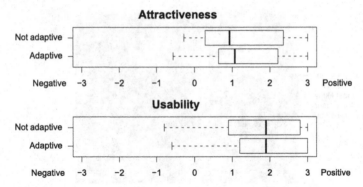

Fig. 6.13. Attractiveness und usability ratings of the "Component Emphasis" adaptation.

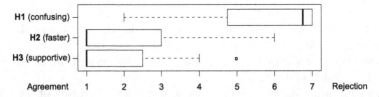

Fig. 6.14. User ratings for the three adaptation hypotheses for the "Component Emphasis" adaptation.

was considered the training task, and the last one was left out because of issues in the user modeling. The average time and number of interactions was computed for the remaining two tasks. In addition, the slowest (V2) and the fastest (V7) user were removed as well as one user (V19), for whom the user modeling did not function.

The results of the objective evaluation are given in Figure 6.15. Both the average time (25.87 seconds for the adaptive vs. 27.47 seconds for the non-adaptive version) and the number of interactions (22.8 interactions in the adaptive vs. 23.8 interactions in the non-adaptive version) only slightly improve with the adaptive version. However, the operating time as well as the number of actions decreases for 71 % (12 out of 17) of the users.

### Discussion and Lessons Learned

Although the "Component Emphasis" adaptation does not improve the perceived attractiveness and usability significantly, the users strongly rate the adaptation as not confusing, faster, and supportive. The number of average interactions and the operating time both slightly decrease. The adaptation improves the objective measures for a majority of 71 % of the users. Therefore, the Component Emphasis presents a viable means to assist users by guiding them to the next interaction.

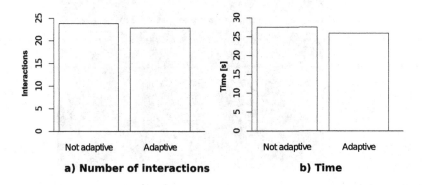

**Fig. 6.15.** Objective measures from the evaluation of the "Component Emphasis" adaptation.

The emphasis of the interface element has to be performed such that the user does not confuse the adaptation with a regular selection. A pre-test version of this test system had two similar visualizations for cursor and emphasis, which were subsequently confused by pre-test users. While this effect was lower with the test version of the system, some users still confused the visualizations in the first test iteration. A number of users suggested putting the selection cursor on the emphasized element. However, other users rejected this idea. In addition, this approach would have inhibited a comparison of the number of user actions in the evaluation.

### 6.4.4 Adaptation Pattern: List Element Selection

The "List Element Selection" adaptation supports the user by highlighting list items. We tested this adaptation with a name selection task in an automotive dashboard interface.

**Test System and Task**

In order to investigate the "List Element Selection" adaptation, an interactive system that resembles an automotive dashboard interface was created. The evaluation task was the selection of names in an address book. The selection screen consists of a list of 65 names in alphabetic order. A scrollbar on the left hand side indicates the current position in the list. The user presses up and down buttons to move the cursor and an OK button to select an item. A screenshot of the system is given in Figure 6.16.

A solid red background and a yellow star highlight a list element in the adaptive version (called "Adaptive"). In this test, the highlight corresponds to the next item that should be selected according to the task. In addition, a small yellow star indicates the position of the highlighted item in the scrollbar.

**Fig. 6.16.** A list of names used in the test system for the "Component Emphasis" adaptation. The system recommends an entry ("Lieb, Sebastian") by highlighting the respective line.

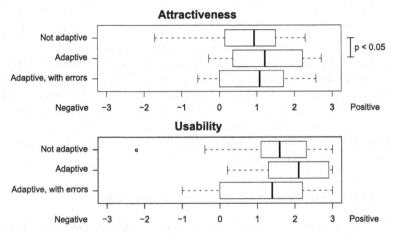

**Fig. 6.17.** Attractiveness ratings of the "List Element Selection" adaptation.

In order to investigate an approach that better reflects the imperfect nature of predictions in real-world situations, a second adaptive version highlights three list items instead of one (named "Adaptive, with errors"). In this case, the adaptation emphasizes the correct item and two wrong items. Only 19 out of 20 subjects performed the evaluation with both adaptive versions. The assignment sheet provided seven names the user had to select in the given order. The number of interactions for all versions was constant to allow a comparison between the different conditions. For this purpose, the sum of the list indices was constant in both conditions.

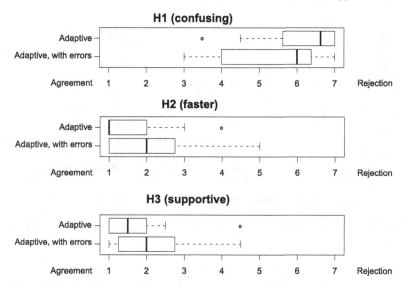

**Fig. 6.18.** User ratings for the three adaptation hypotheses for the "List Element Selection" adaptation.

## Subjective Evaluation

Users rate the attractiveness of the version with the adaptation that highlights one item ("Adaptive") as significantly better ($p \leq 0.05$) and the usability better, yet without significance (see Figure 6.17). Figure 6.18 presents the analysis of the hypotheses: the adaptation is not perceived as confusing (H1, rejection with a median value of 6.50), but as faster (H2, agreement with median value of 1.00) and supportive (H3, agreement with a median value of 1.50). The users did not experience the adaptation as intrusive (rejection with a median value of 6.00). 19 out of 20 subjects (95 %) prefer the adaptive version.

Similar results are observed for the version with three recommendations ("Adaptive, with errors"), but the effect is weaker. The usability of this adaptation is regarded lower, but not significantly. Most users still prefer the adaptive version (15 out of 19), but four prefer the non-adaptive version.

## Objective Evaluation

The slowest (V02) and the fastest subject (V07) were removed for the objective evaluation. In addition, three more subjects (V09, V13, and V17) had to be removed because the subjects talked to the supervisor, who was in the same room. One subject (V08) was excluded due to problems with the user modeling. The analysis was performed with the remaining 14 subjects.

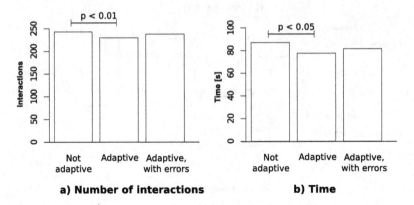

**Fig. 6.19.** Objective measures from the evaluation of the "List Element Selection" adaptation.

Figure 6.19 presents the comparison of the objective measures for the two variants. In the "Adaptive" condition, both the number of clicks ($p \leq 0.01$) and the total interaction time decreased significantly ($p \leq 0.05$). Thus, the adaptation lets users perform the task faster and with less effort. Both time and number of interactions decrease in the "Adaptive, with errors" condition, but without significance.

**Discussion and Lessons Learned**

The "List Element Selection" proved to be a successful adaptation. It was significantly faster than the non-adaptive version and requires a smaller number of interactions. In addition, the users rated the adaptation positively in the questionnaire. The "Adaptive, with errors" version also showed an improvement to the non-adaptive baseline, yet without significance. Therefore, the List Element Selection adaptation presents a viable means to improve the user-system interaction in list selection tasks.

A pre-test version of the test system used a white border for the selection cursor and a red border to indicate the adaptation. However, users were not able to distinguish these two markings. Instead, an additional symbol and a different shape (such as filled rectangle compared to outlined border) allow the user to differentiate the two markings more easily.

### 6.4.5 Adaptation Pattern: Alternative Elements

The "Alternative Elements" adaptation supports the user by selecting the most appropriate interface elements from a number of alternatives. The adaptation was evaluated by means of an automotive dashboard interface.

**Test System and Task**

We selected the task of navigation destination entry for the evaluation of the "Alternative Elements" adaptation. Destination entry comprises the selection of city, street, and house number. In both the non-adaptive and the adaptive version, these values had to be selected from a limited set of 50 values each. The users selected from the 50 largest German cities, the 50 most common street names in Germany, and the values one to 50 as house numbers.

Two slightly different interfaces were created for this system. The "expert mode" leaves more flexibility to the user by offering a menu in which the user may choose the selection order. A screenshot of the main screen of the expert version is given in Figure 6.20. The "beginner mode" enacts a fixed selection order of city first, street next, and house number last. After each value, a confirmation screen ensures that the correct value was selected. Whereas both systems offer the same functionality, the level of control is reduced in the beginner mode. The system is controlled by up and down buttons for scrolling, an OK button for selection, and a back button.

The adaptation automatically selects the appropriate version. The beginner mode is selected at the beginning of each session. Since the subjects are only available for a limited amount of time, the mode is switched to the expert mode after the users had entered two destinations rather than performing actual user modeling. A message indicates the mode selection both of the beginner mode at the beginning and the expert mode after two iterations. Since an investigation of the objective measures would have compared the beginner to the expert mode rather than provide metrics of the adaptation (e.g. with regard to the number of interactions), the objective investigation was omitted for this adaptation.

**Subjective Evaluation**

Since the difference between the two modes was not fully obvious to all users in the beginning of the experiments, the adaptation was again explained before the users filled in the questionnaire of the adaptive condition. The users were instructed not to rate the beginner or the expert mode, but the ability of the system to switch between them.

Both perceived attractiveness (Figure 6.21) and usability (Figure 6.22) improve for the adaptive version, yet not significantly. However, this improvement is even clearer if only users younger than 46 years are investigated. In this case, the improvement of the attractiveness becomes significant ($p < 0.05$). The users' answers to the questions about the three hypotheses are given in Figure 6.23. Users disagree with the statement that the adaptation was confusing, but not strongly (H1, median value of 5.25). They do not regard the adaptation as faster than the non-adaptive version (H2, median value of 4.25). However, they agree that the system was supportive (H3, median value of 3.00). Only small differences can be found between the complete group and

**Fig. 6.20.** The main screen of the expert version of the test system for the "Element Selection" adaptation.

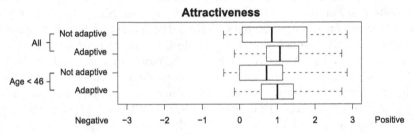

**Fig. 6.21.** Attractiveness ratings of the "Element Selection" adaptation.

the group aged below 46 and therefore, these are not included in the figure. The users do not perceive the adaptation as intrusive (rejection with a median value of 5.00).

In the complete group, 11 out of 20 users (55 %) prefer the adaptive version over the non-adaptive version, seven prefer the non-adaptive version, and two are undecided. However, in the group of subjects aged under 46 years, ten out of 15 subjects (66 %) prefer the adaptive version, three prefer the non-adaptive version, and two are undecided.

### Discussion and Lessons Learned

These findings indicate that users older than 45 years prefer a static interface, whereas younger users like a system that selects the most appropriate version for them. Some older users also stated that once they had learned to use an interface, they did not want to be forced to learn a different one. Younger users on the other hand prefer a dynamic interface.

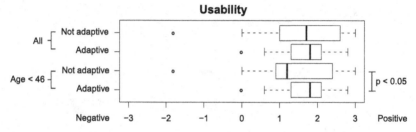

**Fig. 6.22.** Usability ratings of the "Element Selection" adaptation.

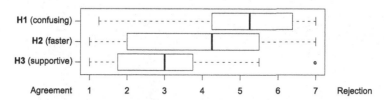

**Fig. 6.23.** User ratings for the three adaptation hypotheses for the "Element Selection" adaptation.

Substantial interface changes should be communicated to the user, for instance by means of explicit messages, in order to avoid confusion. Smaller modifications, such as selecting a different font, may be performed without notification. Further user tests should investigate the replacement of other user interface elements, such as individual graphical elements or speech prompts.

### 6.4.6 Adaptation Pattern: Shortcut Area

The "Shortcut Area" adaptation supports the user by offering specific items, such as list elements or action sequences, in a separate area of the interface. An automotive dashboard interface was used to investigate this adaptation.

### Test System and Task

We evaluated the "Shortcut Area" adaptation with a system that resembles an automotive dashboard interface. It provides access to a "Music" menu, a "Contacts" menu, and a "Climate Control" menu. A screenshot of the main menu is given in Figure 6.24. The "Music" and the "Contacts" menus offer different submenus and selection lists, whereas the "Climate Control" menu allows the user to change the climate settings by means of different buttons. The user controls the interactive system by clicking the buttons on the screen with a mouse, thus resembling a touch screen interface.

The adaptation predicts a sequence of user actions and offers shortcut buttons at the bottom of the screen. The user clicks the buttons to execute

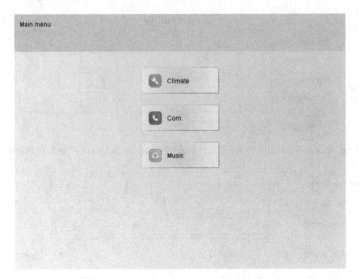

**Fig. 6.24.** The main screen of the test system for the "Shortcut Area" adaptation.

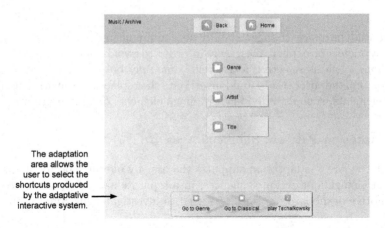

**Fig. 6.25.** A sub-screen of the "Music" menu of the "Shortcut Area" test system. A list of shortcuts is available at the bottom of the screen.

the whole action sequence or a subsequence of it. For example, if the user clicks on the third button, the actions associated with the first, the second, and the third button are executed. Figure 6.25 gives a screenshot of the "Music" menu and the adaptation area on the bottom.

This test system exploits the user modeling algorithm presented in Section 3.4.2. For this purpose, the users received a number of repetitive tasks, such as selecting a certain song from the "Music" menu or changing the climate settings. The users trained the algorithm first by executing four tasks from the assignment sheet. Each task consists of a sequence of actions the

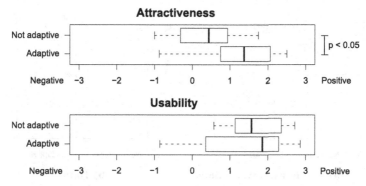

Fig. 6.26. Attractiveness and usability ratings of the "Shortcut Area" adaptation.

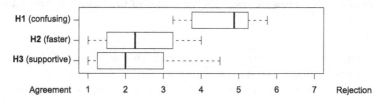

**Fig. 6.27.** User ratings for the three adaptation hypotheses for the "Shortcut Area" adaptation.

users had to follow closely. For instance, one task instructs the user to open the communications menu, open the contacts submenu, and select the name "Martin W.". After the training, the user executed each of the previously trained tasks twice in a given order. The sequence prediction algorithm was trained with user actions to predict the correct sequences by using the same sequences both for training and evaluation. The decision whether to use the adaptations was left to the user's discretion.

This test system was not implemented using the adaptation framework introduced in Chapter 5, since the framework had not been finished when this evaluation was conducted. However, both the user modeling algorithm and the adaptation has been made part of the framework since then. As discussed in the introduction, a different user group than in the previously presented tests performed this evaluation.

**Subjective Evaluation**

Figure 6.26 presents user assessments of attractiveness and usability. Users regard the attractiveness of the adaptive version as significantly better ($p < 0.05$, median values of 1.38 vs. 0.44). While the mean value of the usability rating of the adaptive version (1.86) is slightly better than the value of the

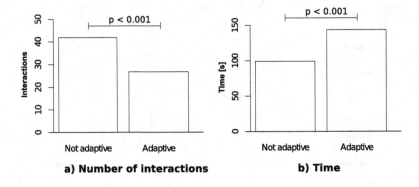

**Fig. 6.28.** Objective measures from the evaluation of the "Shortcut Area" adaptation.

non-adaptive version (1.57), some users rate the usability of the adaptive version as worse, but without significance.

Figure 6.27 shows the results of the three adaptation hypotheses. The subjects do not feel confused, but do not reject this notion strongly (H1, rejection with a median value of 4.88). The users are convinced that the adaptation allows them to use the system faster (H2, agreement with a median value of 2.25) and feel supported by the adaptation (H3, agreement with a median value of 2.00). Ten users prefer the adaptive version, five subjects favor the non-adaptive version or another version, and one person did not select a version.

### Objective Evaluation

The results of the objective evaluation are given in Figure 6.28. Users reduce the overall number of clicks significantly with the adaptive version ($p < 0.001$). Therefore, the users employ the adaptive shortcuts, although they were not forced to do so. However, the interaction time inversely increases significantly. We attribute this to the fact that subjects compared the items in the adaptation area with the next steps on the worksheet.

Thus, users employ the adaptive version by choice, but with a negative impact on the operating time. An evaluation with different tasks or a test with real world problems can reveal if this adaptation also successfully reduces the interaction time.

### Discussion and Lessons Learned

The evaluation of the "Shortcut Area" adaptation shows that subjects employ adaptive shortcuts by choice. A majority of the subjects prefer the adaptive version. The adaptation allows users to reduce the number of interaction steps significantly. However, we attribute an increase in the operating time to the evaluation setup. Further evaluations therefore have to investigate if a different test setup leads to a reduced operating time.

During the evaluation of the "Shortcut Area" adaptation, two additional visualizations were tested. The first one offers only a single button, which executes the whole sequence of five items. The users rejected this version because of the limited flexibility. The second visualization adds a small icon to the interface element that triggers the first action of the predicted sequence instead of using a separate area on the bottom of the screen. The user opens a menu by clicking on the icon to select a sequence of actions to be executed. However, the users rejected this version as well, since it is more complex than other visualizations.

### 6.4.7 Evaluation Summary

We presented an evaluation of interactive systems that employ different adaptations. This evaluation revealed a very positive attitude of users towards adaptive features and a preference of a majority of users for the adaptive version over a non-adaptive baseline version. However, not all differences were significant. Users also comment positively on the specific adaptations. The adaptations also improved the objective measures. A reduction of the number of interactions and the interaction time was observed for most adaptations. Some adaptations only showed a positive effect for a part of the users, such as beginners or older users. In addition to proving the utility of adaptations, this evaluation provides a better insight into the effects of adaptations on the user-system interaction. In doing so, the evaluation contributes to an empirical validation of adaptations and further refines the context of use of the adaptation patterns in Chapter 4. This knowledge may be transferred to other interactive systems that should be extended with adaptivity and thus supports the development process of adaptive interactive systems.

The evaluations were performed on the assumption of a perfect user modeling component that always predicts the correct values. The user modeling component in the evaluation predicted the next value according to the specified task. Since the prediction accuracy of actual algorithms is limited, real-world adaptive interactive systems require an evaluation that investigates adaptations in conjunction with specific user modeling algorithms. For example, Tsandilas and schraefel (2005) present an evaluation of a list selection task that includes prediction accuracy as a variable. Moreover, the tasks used in the evaluation were not real user tasks, but a script that the users followed. An investigation of user interactions with real-world tasks is also needed.

Some adaptations only were successful for a subgroup of the users. For instance, experts did not like the "Adaptive Help Presentation" adaptation because they did not need assistance by the interactive system. Beginners however rated this adaptation very positively. In addition, older users disliked the "Alternative Elements" adaptation, whereas younger users accepted this adaptation. Therefore, some adaptations should be enabled for specific user groups. In addition, adaptive systems should allow the user to disable the adaptations.

Adaptations should be employed in interactive systems in a prudent way. The adaptations presented in this chapter may serve as a guideline for how to design the adaptations. The mere use of adaptations does not improve the user-system interaction: Early test versions of some test systems used for the evaluation implemented adaptations such that they confused test users or the adaptations were not even noticed by the subjects. By taking these findings into account, the adaptations could be implemented in a successful way for the final test systems. Thus, adaptations have to be implemented according to findings from evaluations. Moreover, adaptations should be employed such that they support the appropriate user group, such as beginners or older users.

## 6.5 Discussion

In this chapter, we presented an evaluation of the approaches presented in previous chapters, namely user modeling algorithms and a set of adaptation patterns. A review of literature suggested a separate evaluation of the individual components of an adaptive interactive system. Therefore, we examined the components of adaptive interactive systems separately. First, an investigation examined attitudes towards adaptivity of two user groups, one of users of adaptive interactive systems and a larger one of international users. These users show a positive attitude towards adaptive features in interactive systems. The user modeling algorithms presented in Chapter 3 were tested with recorded log data. First, an evaluation of the PDFA-based action recognition algorithm investigated the results in relation to the number of training sessions and showed good recognition rates already with a small number of sessions. Second, an evaluation of two action prediction algorithms, namely Markov-based and task model prediction, showed a better performance of the Markov-based algorithm. A user test with five different adaptive interactive systems examined the adaptations introduced in Chapter 4. In most cases, the adaptations showed an improvement both in the interaction time and the number of interactions, although not always with significance. In doing so, the evaluation refines the context of use of these adaptations. Although the evaluation was based on strict task assignments, adaptations proved to be a viable means for improving the usability of interactive systems. Further tests should investigate the situations in which adaptation had a negative impact on the user-system interaction. A questionnaire collected users' attitude towards adaptations and revealed that adaptations improve the perceived quality of an interactive systems in addition to objective measures. However, some adaptations only worked for a subgroup of the users, such as beginners or older users.

Thus, we demonstrated the feasibility of the adaptation approach developed in this work with user tests, including the user modeling algorithms and the adaptation patterns. In addition, we showed the viability of the adaptation framework with a reference implementation by applying it to different interaction systems. Adaptations should be implemented in a prudent way and according to evaluation evidence. Otherwise, adaptations affect the usability negatively. Pre-test versions revealed problems that were addressed before the evaluation. The solutions were discussed in the respective sections. However, the evidence from this evaluation should be expanded and further strengthened as future work. In addition to evaluating more interactive systems, the evaluation should address adaptive speech-based dialog systems. We discussed the application of the adaptations to voice interfaces in the pattern descriptions. In order to support a widespread use of adaptations, empirical evidence on user modeling algorithms and the use of adaptations has to be made available to developers. This chapter contributes to building such a body of evidence. Developers may consult the findings presented in this chapter when designing adaptive applications.

# 7

# Summary and Outlook

I like the dreams of the future better than the history of the past.

–Thomas Jefferson (1743–1826)

In this book, we discussed the adaptation of multimodal interactive systems to user behavior. The building blocks of the approach are a user modeling component, a set of adaptation patterns, and an adaptation framework. The user modeling component describes user behavior from basic events. Based on a recognition of user actions, a higher-level description of user behavior and a prediction of actions and preferences become feasible. For the adaptation description, we collected a set of adaptations and and documented them as patterns in a specific format. The adaptation framework integrates the user modeling approach and the adaptations. A reference implementation of this architecture shows the practicability of the presented approaches and serves as a test bed for an evaluation. We performed an evaluation with test subjects to investigate both the user modeling algorithms and the presented adaptations. In the remainder of this chapter, we give a summary of this work and give an outlook to future research.

## 7.1 Summary

In this section, we summarize the chapters of this book, which discuss a set of user modeling algorithms, multimodal adaptation patterns for interactive systems, and an adaptation framework. After a general introduction to adaptive interactive systems in Chapter 1, we discussed related work for the individual topics in Chapter 2.

User modeling describes the observation of the user-system interaction. We regard user behavior as a sequence of basic events. The user modeling procedure only considers user input and system reactions that are represented by such an event. Different modalities, such as haptic and speech input, produce these basic events. User behavior comprises actions, which are meaningful sequences of events, and data, which further defines these actions. For example, if a user changes the channel to "BBC" in a TV system, the user performs an action called "ChangeChannel" and the channel name is associated with the action as data. In Chapter 3, we presented an approach we devised for detecting user actions in a sequence of events. This approach uses probabilistic automata to recognize different sequences that describe a specific action. Once user actions have been detected, a description of higher-level user behavior becomes feasible. We introduced task models as a technique for defining possible user actions in adaptive interactive systems. Approaches in the literature mostly apply the task model at design time. Instead, we apply the task model at runtime of an interactive system. Different adaptation triggers may be derived from these models. We present different kinds of information that may be extracted from a task model, such as recommending unused actions or detecting problems in the user-system interaction. We present how this information serves as an adaptation trigger. Thereafter, we presented different action prediction algorithms. First, a prediction is derived from a task model that was enriched with statistical information. Second, Markov chains are used for modeling action sequences and predicting the next action. This algorithm is adapted to interactive systems from an existing prediction algorithm. Domain knowledge is added to the statistical algorithm to improve the prediction. An evaluation shows the feasibility of the approaches, compares the two algorithms, and exposes the individual advantages. In addition, we introduce an algorithm for predicting sequences of actions. This algorithm employs sequence mining for computing the most likely action sequence to follow. These algorithms are tailored to interactive systems.

The outcomes of the user modeling process trigger adaptations. Since usability is an important issue for adaptations, we started Chapter 4 with a review of usability principles for interactive systems and discussed their implications for adaptive interfaces. We introduced design patterns as a technique for communicating best practice in a domain. Thereafter, we presented a pattern format we created for adaptation patterns in interactive systems. Finally, we defined a set of adaptation patterns, which we derived from a review of literature and from adaptations we implemented for this work. The adaptation patterns are applicable both to graphical and speech interfaces. Developers may consult these reusable patterns when adding adaptive features to multimodal user interfaces. Thus, these patterns simplify the task of selecting appropriate adaptations for interactive systems.

In Chapter 5, we presented an adaptation framework that covers the complete process from an observation of basic events to the execution of adaptations. This framework allows system designers to integrate adaptations more easily into both graphical and speech interfaces. The framework comprises three main constituents: a semantic layer, a user modeling component, and an adaptation component. The semantic layer employs an ontology to create a common representation of the interactive system and other topics, such as the user and the domain. In the user modeling component, basic events connect different components that perform user modeling. Higher-level models, such as an interaction model and a task model, perform additional inferences. A user model is connected to basic events and updates user model entries from these events. In addition, the user model performs further computations with different user modeling algorithms, such as computing preferences or predicting actions. An adaptation component divides the definition of adaptations into a system-independent and a system-dependent part. In doing so, adaptations may be reused between different interactive systems and at the same time be fitted to the requirements of a specific system. A reference implementation of the framework serves as a test bed for an evaluation.

In order to show the practicability of these approaches and to provide empirical evidence for adaptive interactive systems, we conducted user tests with different interactive systems. The results of these tests were presented in Chapter 6. An evaluation of the user modeling algorithms proves the feasibility of the action recognition algorithm. In addition, a comparison of the two action prediction algorithms revealed that the Markov chain algorithm produces better predictions than the task model algorithm, yet not significantly. In order to investigate the different adaptations, we implemented a separate interactive system for each adaptation. In a user test, subjects were handed an assignment sheet with tasks they followed closely. A questionnaire collected users' attitudes towards the individual adaptations. In general, the subjects showed a very positive attitude towards the concept of adaptivity. Moreover, objective measures, namely the interaction time and the number of interaction steps, were improved by the adaptations. However, some adaptations should only be applied for specific user groups. For instance, some adaptations were more successful for beginners or younger users and did not work for experts or older users. Since the individual components have been investigated separately, these results may be transferred to other interactive systems. Thus, we provide evaluation evidence for adaptations in interactive systems.

Thus, adaptations proved to be a feasible approach for improving the usability and attractiveness of interactive systems. We demonstrated the viability of the presented user modeling algorithms and adaptations in a user test. In addition, we illustrated the feasibility of the presented framework by implementing different test systems with the framework. In doing so, we presented solutions in this work for the individual problems identified by Höök (2000) that have to be addressed to allow a widespread use of adaptive interfaces, namely usability, useful adaptations, development methods, and maintainability. Therefore, this framework, in conjunction with the presented algorithms and adaptations, presents a viable foundation for a use of adaptations in interactive systems.

## 7.2 Outlook

In this section, we discuss research that may be conducted in the domain of adaptive interactive systems. First, the number of adaptive interactive systems developed with this framework may be increased. The adaptation patterns presented in Chapter 4 are generic and more concrete adaptations may be derived from these patterns. For this purpose, the list of adaptation selectors and executors that are part of the framework may be extended. Although we performed a thorough investigation of existing adaptations, new patterns might possibly be identified.

In addition, a detailed study and evaluation of adaptive speech interfaces may be performed. The adaptation patterns discuss multimodal interactive systems and include adaptations for speech interfaces. However, an evaluation of these interfaces may be performed to provide empirical evidence for speech dialog systems. In addition, more adaptation executors and selectors for speech interfaces further improve the scope of the framework.

Moreover, the automatic strategy of the adaptation component may be refined. Currently, the adaptation component employs a strategy that activates adaptations based on the number of active adaptations and the experience of the user. This strategy may be improved by enabling the adaptation component to learn an optimal strategy for applying adaptations for an individual user. For this purpose, the adaptation component may be equipped with self-assessment capabilities to determine if the applied adaptations have been successful.

In addition to adapting to user behavior, an interactive system may integrate other adaptation causes. For example, an interface may adapt to the location of the user by reading the GPS position or to the time of the day by reading a clock. These types of information may be integrated into the framework by emitting GPS or time events. After user modeling has been performed, these causes may be connected to the adaptation component.

The adaptation framework presented in Chapter 5 builds on a model-based development environment. The foundation of the framework may be strengthened by using a formal foundation, such as UML. In doing so, all development tools that are based on UML may employ the adaptation framework. Formal models transformations could be used between the different formalisms. Different reference models exist in the domain of adaptive hypertext. Therefore, the ontology used by the semantic layer may be extended into a reference model.

# A

# Adaptation Patterns for Interactive Systems

In this chapter, we present the adaptation patterns introduced in Section 4.3 in the pattern format introduced in Section 4.2.

## A.1 Component Emphasis

### Intent

Guide the user by emphasizing certain elements of the interface. Limit the changes to the part of the interface that requires emphasis. In doing so, enable users to reuse acquired knowledge of the interactive system and avoid distracting the user through fundamental changes of the interface.

### Motivation

During the interaction with an interactive system, a user has a goal and is looking for interface elements that helps in fulfilling it. For instance, the user may look for a graphical button to trigger an action and the system provides support by guiding the user to the respective interface element.

### Forces

- The user follows a certain goal when using the interactive system and may spent considerable time looking for interface elements that facilitate reaching this goal.
- Performing major changes to the system confuse the user and distract from the current task. Subtle guidance instead supports the user.
- Since emphasizing wrong elements impedes the user, an appropriate user modeling prediction is crucial for this adaptation.
- The adaptive emphasis should be conceived in a way that the user does not confuse it with a regular selection in the user interface.

## Solution

Make the adaptive system change properties of interface elements in a way as to draw the user's attention. Use assumptions of a user modeling component, such as a prediction of the most likely next action or an action the user has not used yet. Help the user reach the current goal by emphasizing interface elements that are related to the respective assumption of the user modeling component.

## Adaptation Trigger

The following observations of the user modeling component trigger the Component Emphasis adaptation:

- Prediction of the next user action.
- Actions that the user has not used yet, but which others have used.

## Related Patterns

The "List Element Selection" pattern emphasizes elements in a list and is therefore related to this pattern, which emphasizes arbitrary elements that are related to user actions.

The "Prominent done button" pattern (Tidwell, 2005) statically emphasizes a button that finishes a task associated with a graphical view, but the emphasis is not performed based on the current user's behavior.

## Example

Consider an electronic program guide, in which the user specifies filter criteria, such as channel or time, to filter the list of TV shows. After a number of criteria were selected, the user has to press a "Show results" button to see all shows that match the selected criteria. Increasing the size of the button and changing colors (see Figure A.1) emphasizes the button, thus supporting the user in finishing the current task.

The Component Emphasis pattern is also applicable to voice interfaces. If a user enters a state in which the system reads the list of possible utterances, saying a phrase as the first or the last one draws a user's attention to this phrase.

# A.2 List Element Selection

## Intent

Support the user in selecting elements from a list, for instance by highlighting frequently used entries from the list.

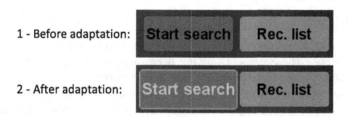

1 - Before adaptation:   Start search    Rec. list

2 - After adaptation:    Start search    Rec. list

**Fig. A.1.** Emphasis of a button in an interactive TV system. The "Start search" button is emphasized compared to the non-emphasized "Rec. list" button. The reasoning is to provide non-intrusive and subtle hints, in this case by increasing the size of the button and changing the text color.

## Motivation

When selecting elements from a list, users select some elements frequently and others not at all. The selection process is improved by emphasizing frequently selected elements from the list.

## Forces

- Selecting frequently used items in a list should take less time for the user than selecting others.
- If a list is longer than one screen, highlight the interesting items also in the scrollbar to enable the user to quickly scroll to the interesting elements.
- Emphasized list elements should be highlighted in a way that the user does not confuse the emphasis with a selection cursor.
- Since emphasizing wrong elements impedes the user, a sufficiently good user modeling prediction is crucial for this adaptation.

## Solution

Emphasize these elements in the list that have been selected more often before than others. In doing so, let the user see more quickly elements that are of increased interest. For instance, change the text or background color of these elements or add markers to differentiate interesting elements from others.

## Adaptation Trigger

The following observations of the user modeling component trigger the List Element Selection adaptation:

- List entries that have been selected more often than others either by the current user or by other users.
- Elements in a list that the user has not yet selected, but which should be interesting based on the user's previous behavior.

**Related Patterns**

The "Element Emphasis" adaptation pattern also emphasizes interface elements. However, these elements are not necessarily similar, as are list elements, and are mostly used for navigating within the system.

The "Annotated scrollbar" pattern (Tidwell, 2005) recommends adding information to the scrollbar, which is also proposed by this pattern to mark the position of recommended elements in the list. Moreover, the "Adaptive Anchor Annotation" (Koch and Rossi, 2002) pattern describes how to annotate links in a hypertext system, for instance by adding the level of interest for the user.

**Example**

Selecting elements from a list is a very common task when interacting with interactive systems. For instance, selecting a name from the address book is one of the fundamental functions of interactive systems that support phone calls, such as mobile phones or automotive dashboard systems. Since users call a small number of people from their phone book frequently, the selection of these names from the address book is improved by highlighting these names. An example of such a system is given in Figure A.2, which shows an address book that emphasizes the three most frequently selected elements.

Different visualizations of the "List Element Selection" pattern are possible and have been examined by research projects. One example is the fisheye menu (Bederson, 2000) that assigns a different font size to different elements; the fisheye visualization can be employed for adaptations as well. Voice interfaces emphasize list elements by reading interesting elements first or by adding acoustic markings.

**Fig. A.2.** Three elements are emphasized in a selection list by the List Element Selection pattern to support the selection of frequently elements. Only one item is visible in the current screen and the positions of the other elements are shown in the scrollbar.

# A.3 Alternative Elements

## Intent

Provide a set of configurations for different interface components or screens and select the appropriate configuration to better support the needs of an individual user.

## Motivation

Since the requirements as well as the skills of users of interactive systems vary, some system configurations better reflect the needs of an individual user. Instead of providing one configuration that tries to consider all users, the adaptation selects the version that is best suited for the needs of the current user. A user modeling component provides information about characteristics of the user, such as the proficiency with interactive systems.

## Forces

- Different configurations of interface components or graphical screens better reflect the needs of individual users.
- Automatically generated alternatives can break with existing usability principles.
- The developer has to spend additional time developing the different alternatives, but the user benefits from an improved user-system interaction. In addition, annotation has to be provided that allows the adaptation to select the most appropriate alternative.

## Solution

Provide different versions of a specific part or component of the interactive system to the adaptation component, for instance of a graphical screen, a speech output prompt, or a property (e.g. font size). Support users by selecting the appropriate alternative for the respective element using information from the user modeling component. By providing a set of alternatives to the adaptation component, which were created by the system designer, all variants of the interactive system adhere to design principles.

## Adaptation Trigger

The following observations of the user modeling component trigger the Alternative Elements adaptation:

- Preferences or properties of the user, such as the knowledge level or experience of the user.

**Related Patterns**

The "Alternative views" pattern (Tidwell, 2005) lets the user decide among alternative views, for instance of a web page. However, the most appropriate view is not selected automatically.

The "Adaptive Conditional Fragments" pattern (Koch and Rossi, 2002) personalizes the content of text fragments in adaptive hypertext systems by selecting the most appropriate one of a set of alternative content nodes.

**Example**

The Alternative Elements pattern can be employed at different levels. For instance, if a user has to enter different values in an input screen, such as selecting the destination in a navigation device or selecting criteria in an interactive TV program guide, a simple version of the screen is provided to novice users and a more powerful version to advanced users. On a lower level, a larger font size improves the readability for visually impaired users.

On the other hand, a speech interface can provide different levels of speech output prompts. Novice users receive extended prompts that explain the most important functions when they enter a new part of the system. Intermediate users only require shorter prompts, which list the commands, but do not necessarily explain them. Finally, expert users, who could be annoyed by long and repetitive speech output, only hear a short prompt explaining the current state of the system and receive more explanation only on request. A system that employs this kind of adaptation is for example presented by Hassel and Hagen (2006). Other speech features for this pattern are dialog initiative (user or system initiative) or confirmation style (explicit, implicit, or none).

## A.4 Adaptive Help Presentation

**Intent**

Present adaptive help for the user's current situation.

**Motivation**

Help in interactive systems is often static or only considers the currently active screen, but different people are likely to have different problems in different contexts. Providing help to the user is more valuable if it covers the current task of the user. By taking into account not only the current context, i.e., the graphical screen or speech state, but also the user's interaction history and other user characteristics, help is more specific and thus supports a user more precisely in the current task.

## Forces

- Help tailored to the current task of the user is more valuable than static help.
- Static help can be too advanced for beginners and at the same time too superficial for expert users.
- Providing help can be assistive for beginners, but annoying for expert users.

## Solution

Provide specific help for the current situation of the user. Observe the user-system interaction to determine the situation and the context of the user. Present the help either on a separate area of the screen, or use an icon (or a sound) to indicate the availability of help. For instance, give the user an option to open this help once it is available. However, ensure at the same time that the help does not distract the user. Therefore, avoid messages that fully engage the user's attention, as for instance modal help messages. Provide an acoustic signal instead of a graphical hint for speech interfaces or visually impaired users.

## Adaptation Trigger

The following observations of the user modeling component trigger the Adaptive Help Presentation adaptation:

- Prediction of the next user action.
- Detection of user problems.
- Preferences or properties of the user, such as the knowledge level or experience of the user.

## Related Patterns

The "Multi-level help" pattern (Tidwell, 2005) suggests using different help techniques. Adaptive help is one kind of help that provides information adjusted to the characteristics and current situation of the user.

## Example

In an interactive TV system, the user can browse the TV program in an electronic program guide and for this purpose specify different filter criteria, such as channel or time. Help is presented to the user by fading in a yellow message box on the top of the screen. When the user enters the selection screen for the first time, the help explains how to select filter criteria. After some criteria were selected, the help text on the screen tells the user to open the result screen next. For this purpose, the interaction history is considered. Figure A.3 gives an example of the adaptive help feature in a digital TV system.

Dix et al. (2004) differentiate active help, which presents help messages to the user automatically, and passive help, in which the user request the help. Adaptive help is most useful if it does not distract or annoy the user, but still supports the user and introduces new features.

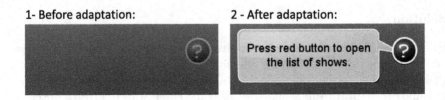

**Fig. A.3.** Adaptive help supports the user by showing a text message that fits the current situation of the user. If a user is sufficiently proficient in working with the system, help messages are no longer shown.

## A.5 Shortcut Area

### Intent

Present shortcuts for executing actions or selecting values to the user on a separate part of the interface. In doing so, accelerate the execution of frequent actions or sequences of actions and selection of the user's favorite values.

### Motivation

Users often perform some action sequences frequently, such as selecting specific elements from a list (e.g. a list of fonts), or by executing the same actions over and over again. The interactive system supports the user by presenting these items as shortcuts. By employing a special area for the shortcuts, the decision whether to use shortcuts is left to the user.

### Forces

- Finding frequently used elements and executing actions repeatedly can be very time-consuming for the user. Shortcuts can therefore simplify the user-system interaction.

- Shortcuts that automatically pop up on top of the interface interfere with the user interface and distract the user. A separate area that is always visible instead allows the user to decide whether or not to use shortcuts and limits the distraction of the user.
- The shortcuts may either be different alternatives, such as different fonts to select from, or a sequence of items, from which the user may pick a subsequence, such as in a sequence of actions.

## Solution

Employ a separate area of the screen – called shortcut area – to present shortcuts to the user, thus avoiding a distraction of the user. Make this list either part of one interface element (e.g. of a list) or make it a separate part of the whole screen for presenting global shortcuts. In doing so, enable the user to find frequently used elements more quickly by selecting them from a distinct area of the screen. Use the output of a user modeling component to create the list of shortcuts.

## Adaptation Trigger

The following observations of the user modeling component trigger the Shortcut Area adaptation:

- Prediction of the next user action or a sequence of user actions.
- Prediction of a user preference, such as a TV channel.

## Related Patterns

The "Streamlined repetition" pattern (Tidwell, 2005) suggests considering repeated operations when creating an interface. The Shortcut Area provides an automatic solution for this recommendation. The "Action panel" pattern (Tidwell, 2005) presents a list of available actions to the user, which is similar to a Shortcut Area that contains user actions. If the Shortcut Area presents a sequence instead of single items, the adaptation is similar to the "Autocomplete" pattern (Tidwell, 2005), since the adaptation anticipates user behavior.

## Example

In a selection list, a separate area on the top of the list presents the most frequently selected entries of the list to the user. By selecting them, the user does not have to scroll through the whole list. One example of such a list is the font selection list in Microsoft Word (2000 and later), which shows the most recently used fonts in a separate area on the top of the list.

A different application of the Shortcut Area pattern is to provide navigation shortcuts. A user modeling component recognizes user actions and predicts a sequence of possible next actions, with each action being represented
by a button in an adaptation area. If the user presses one of these buttons, the
action associated with the button and all actions before the pressed button
are executed, thus reducing the number of required interactions. An example of an interface that provides navigation shortcuts to the user is shown in
Figure A.4: In addition to the regular interface (shown on top), the interactive system presents a list of likely next actions to the user on the bottom
of the screen. If the user presses one of these buttons, the interactive system
automatically executes the respective actions.

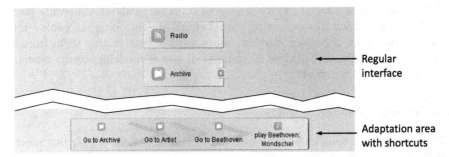

**Fig. A.4.** A separate adaptation area presents a list of buttons based on a prediction
of the user's next actions. These actions are executed by pressing the respective
buttons.

# B

## The Evaluation Questionnaire

In the following, the individual pages of the questionnaire from the user evaluation are presented. Whereas the original questionnaire is in German, the one presented in this chapter was translated into English. The evaluation presented in Section 6.4 compares a non-adaptive and an adaptive version of five interactive test systems. For this purpose, each of the test subjects performed a number of tasks with both versions of the interactive systems.

The users filled in Questionnaire V1 after the non-adaptive version and Questionnaire V2, which consists of two pages, after the adaptive version. Questionnaire V1 contains a block of attribute pairs to poll users' attitudes towards the interactive system. These pairs are based on the AttrakDiff questionnaire Hassenzahl et al. (2003). In addition, users are asked to select whether they prefer the adaptive version, the non-adaptive version, or if they are undecided. This question should only be filled in after the users have tested both versions. Finally, a text box allows users to enter general comments. Questionnaire V2 corresponds to V1, but adds a number of questions about the adaptive features of the tested system. After having tested both versions, the users filled in Questionnaire G with questions about general attitudes towards adaptations and a set of questions about personal data for a demographic analysis.

## Questionnaire V1

| Variant: „non-adaptive" | User: | System: |
|---|---|---|

**1 – Assessment of the system**

*Please state your impression of the interactive system you tested by means of the following pairs of words:*

| complicated | □ | □ | □ | □ | □ | □ | □ | simple |
|---|---|---|---|---|---|---|---|---|
| pleasant | □ | □ | □ | □ | □ | □ | □ | unpleasant |
| practical | □ | □ | □ | □ | □ | □ | □ | impractical |
| ugly | □ | □ | □ | □ | □ | □ | □ | beautiful |
| indirect | □ | □ | □ | □ | □ | □ | □ | direct |
| likeable | □ | □ | □ | □ | □ | □ | □ | unlikeable |
| predictable | □ | □ | □ | □ | □ | □ | □ | unpredictable |
| rejecting | □ | □ | □ | □ | □ | □ | □ | inviting |
| good | □ | □ | □ | □ | □ | □ | □ | bad |
| repulsive | □ | □ | □ | □ | □ | □ | □ | attractive |
| confusing | □ | □ | □ | □ | □ | □ | □ | clear |
| motivating | □ | □ | □ | □ | □ | □ | □ | daunting |

**2 – Comparison of the two variants**

**-- Please only fill in this question if you have used both the adaptive and the non-adaptive version of the test system. --**

I like the following version better:

| □ Adaptive | □ **Non**-adaptive | □ Don't know |
|---|---|---|

**3 – General comments**

*If you have further comments, please write them down in the following text box. You may leave the text box empty.*

## Questionnaire V2 (1/2)

| Variant: „adaptive" | User: | System: |
|---|---|---|

### 1 – Assessment of the system

*Please state your impression of the interactive system you tested by means of the following pairs of words:*

| | | | | | | | | |
|---|---|---|---|---|---|---|---|---|
| complicated | ☐ | ☐ | ☐ | ☐ | ☐ | ☐ | ☐ | simple |
| pleasant | ☐ | ☐ | ☐ | ☐ | ☐ | ☐ | ☐ | unpleasant |
| practical | ☐ | ☐ | ☐ | ☐ | ☐ | ☐ | ☐ | impractical |
| ugly | ☐ | ☐ | ☐ | ☐ | ☐ | ☐ | ☐ | beautiful |
| indirect | ☐ | ☐ | ☐ | ☐ | ☐ | ☐ | ☐ | direct |
| likeable | ☐ | ☐ | ☐ | ☐ | ☐ | ☐ | ☐ | unlikeable |
| predictable | ☐ | ☐ | ☐ | ☐ | ☐ | ☐ | ☐ | unpredictable |
| rejecting | ☐ | ☐ | ☐ | ☐ | ☐ | ☐ | ☐ | inviting |
| good | ☐ | ☐ | ☐ | ☐ | ☐ | ☐ | ☐ | bad |
| repulsive | ☐ | ☐ | ☐ | ☐ | ☐ | ☐ | ☐ | attractive |
| confusing | ☐ | ☐ | ☐ | ☐ | ☐ | ☐ | ☐ | clear |
| motivating | ☐ | ☐ | ☐ | ☐ | ☐ | ☐ | ☐ | daunting |

### 2 – Comparison of the two variants

**-- Please only fill in this question if you have used both the adaptive and the non-adaptive version of the test system. --**

I like the following version better:

| ☐ | ☐ | ☐ |
|---|---|---|
| Adaptive | **Non**-adaptive | Don't know |

**-- Please turn over – more questions on the following page! --**

## Questionnaire V2 (2/2)

### 3 – Questions about the adaptation

*Please rate the adaptation in the interactive system you just used by means of the following statements:*

| | | | | | | | | | |
|---|---|---|---|---|---|---|---|---|---|
| The adaptation distracts me from the current goal. | true | ☐ | ☐ | ☐ | ☐ | ☐ | ☐ | ☐ | not true |
| I can use the system more quickly due to the adaptation. | true | ☐ | ☐ | ☐ | ☐ | ☐ | ☐ | ☐ | not true |
| I can use the system more easily thanks to the adaptation. | true | ☐ | ☐ | ☐ | ☐ | ☐ | ☐ | ☐ | not true |
| The adaptation supports me when using the system. | true | ☐ | ☐ | ☐ | ☐ | ☐ | ☐ | ☐ | not true |
| The adaptation confuses me. | true | ☐ | ☐ | ☐ | ☐ | ☐ | ☐ | ☐ | not true |
| The adaptation prolongs the time I need for using the system. | true | ☐ | ☐ | ☐ | ☐ | ☐ | ☐ | ☐ | not true |
| The adaptation enables me to execute the actions better. | true | ☐ | ☐ | ☐ | ☐ | ☐ | ☐ | ☐ | not true |
| I perceived the adaptation as intrusive. | true | ☐ | ☐ | ☐ | ☐ | ☐ | ☐ | ☐ | not true |
| The adaptation obstructed my interaction with the system. | true | ☐ | ☐ | ☐ | ☐ | ☐ | ☐ | ☐ | not true |

### 3 – General comments

*If you have further comments, please write them down in the following text box. You may leave the text box empty.*

# Questionnaire G

| Variant: „General" | User: |
|---|---|

## 1 – Attitudes towards adaptations

*Please state your general attitude towards adaptations by means of the following statements.*

| | | | | | | | | | |
|---|---|---|---|---|---|---|---|---|---|
| In general, I like the presented adaptations. | true | ☐ | ☐ | ☐ | ☐ | ☐ | ☐ | ☐ | not true |
| I think adaptation is an interesting concept. | true | ☐ | ☐ | ☐ | ☐ | ☐ | ☐ | ☐ | not true |
| I think adaptation will be used commercially in the future. | true | ☐ | ☐ | ☐ | ☐ | ☐ | ☐ | ☐ | not true |
| If I could choose between an adaptive and a non-adaptive system, I would choose the **non-adaptive** system. | true | ☐ | ☐ | ☐ | ☐ | ☐ | ☐ | ☐ | not true |
| If I could choose between an adaptive and a non-adaptive system, I would relinquish other features in favor of adaptations. | true | ☐ | ☐ | ☐ | ☐ | ☐ | ☐ | ☐ | not true |

## 2 – Personal data

*Finally, we ask you to provide some personal data. This data does not allow conclusions about you and the data is only collected to allow a demographic analysis of the evaluation.*

| | | | | | | | |
|---|---|---|---|---|---|---|---|
| Gender | | ☐ male | | | ☐ female | | |
| Age | ☐ < 18 | ☐ 18-25 | ☐ 26-35 | ☐ 36-45 | ☐ 46-55 | ☐ 56-65 | ☐ > 65 |
| I'm a computer... | ☐ beginner | | ☐ advanced user | | ☐ expert | | |

## 3 – General comments

*If you have further comments, please write them down in the following text box. You may leave the text box empty.*

# List of Figures

# List of Tables

# References

G. Adomavicius and E. Tuzhilin. Toward the Next Generation of Recommender Systems: A Survey of the State-of-the-art and Possible Extensions. *IEEE Transactions on Knowledge and Data Engineering*, 17:734–749, 2005.

R. Agrawal and R. Srikant. Mining Sequential Patterns. In *International Conference on Data Engineering (ICDE) 1995*, pages 3–14. IEEE Computer Society, Washington, DC, USA, 1995.

C. Alexander, S. Ishikawa, and M. Silverstein. *A Pattern Language. Towns, Buildings, Construction.* Oxford University Press, New York, NY, USA, 1977.

S. Amershi and C. Conati. Unsupervised and Supervised Machine Learning in User Modeling for Intelligent Learning Environments. In *Intelligent User Interfaces (IUI) 2007*, pages 72–81. ACM, New York, NY, USA, 2007.

G. Amores, G. Pérez, P. M. Portillo, F. Gómez, and J. González. Integrating OWL Ontologies with a Dialogue Manager. *Procesamiento del Lenguaje Natural*, 37:153–160, 2006.

J. Annett and K. D. Duncan. Task Analysis and Training Design. *Journal of Occupational Psychology*, 41:211–221, 1967.

D. E. Appelt. Introduction to Information Extraction. *AI Communications*, 12(3):161–172, 1999.

Apple Inc. Apple Human Interface Guidelines. Online, 2010. URL http://developer.apple.com/Mac/library/documentation/UserExperience/Conceptual/AppleHIGuidelines/. Retrieved 2010-04-07.

A. Aragones, J. Bruno, A. Crapo, and M. Garbiras. An Ontology-Based Architecture for Adaptive Work-Centered User Interface Technology. Technical report, General Electric Inc., 2007.

L. Ardissono, C. Gena, P. Torasso, F. Bellifemine, A. Chiarotto, A. Difino, and B. Negro. Generation of Personalized Electronic Program Guides. In *Advances in Artificial Intelligence (AI*IA) 2003*, volume 2829 of *Lecture Notes in Computer Science*, pages 474–486. Springer Verlag, Heidelberg, Germany, 2003.

L. Ardissono, C. Gena, P. Torasso, F. Bellifemine, A. Difino, and B. Negro. *Personalized Digital Television*, volume 6 of *Human-Computer Interaction Series*, chapter User Modeling and Recommendation Techniques for Personalized Electronic Program Guides, pages 3–26. Springer, Dordrecht, The Netherlands, 2004.

K. Bachfischer, T. Bohnenberger, M. Hofmann, C. Wäller, and Y. Wu. Kontext-adaptive Fahrerinformationssysteme am Beispiel eines Navigationssystems. *Künstliche Intelligenz*, 3:57–63, 2007.

L. K. Baumeister, B. E. John, and M. D. Byrne. A Comparison of Tools for Building GOMS Models. In *Conference on Human Factors in Computing Systems (CHI) 2000*, pages 502–509. ACM, New York, NY, USA, 2000.

T. Becker, P. Poller, J. Schehl, N. Blaylock, C. Gerstenberger, and I. Kruijff-Korbayov. The SAMMIE System: Multimodal In-car Dialogue. In *Conference on Computational Linguistics (Coling ACL) 2006 on Interactive presentations*, pages 57–60. Association for Computational Linguistics, Morristown, NJ, USA, 2006.

B. B. Bederson. Fisheye Menus. In *User Interface Software and Technology (UIST) 2000*, pages 217–225. ACM, New York, NY, USA, 2000.

P. V. Biron and A. Malhotra. XML Schema Part 2: Datatypes Second Edition. W3C recommendation, World Wide Web Consortium, 2004. URL http://www.w3.org/TR/xmlschema-2/. Retrieved 2010-10-08.

R. A. Bolt. "Put-that-there": Voice and Gesture at the Graphics Interface. In *Conference on Computer Graphics and Interactive Techniques (SIGGRAPH) 1980*, pages 262–270. ACM, New York, NY, USA, 1980.

J. Borchers. *A Pattern Approach to Interaction Design*. Wiley, Chichester, UK, 2001.

J. Breese, D. Heckerman, and C. Kadie. Empirical Analysis of Predictive Algorithms for Collaborative Filtering. In *Conference on Uncertainty in Artificial Intelligence (UAI) 1998*, pages 43–52. Morgan Kaufmann, Madison, WI, USA, 1998.

M. Broy. Challenges in Automotive Software Engineering. In *International Conference on Software Engineering (ICSE) 2006*, pages 33–42. ACM, New York, NY, USA, 2006.

P. Brusilovsky. Adaptive Hypermedia. *User Modeling and User-Adapted Interaction*, 11(1-2):87–110, 2001.

P. Brusilovsky, C. Karagianidis, and D. Sampson. The Benefits of Layered Evaluation of Adaptive Applications and Services. In *Workshop on Empirical Evaluation of Adaptive Systems (EASy) 2001*, pages 1–8. Pedagogical University of Freiburg, Freiburg, Germany, 2001.

P. Buitelaar. OntoSelect: Towards the Integration of an Ontology Library, Ontology Selection and Knowledge Markup (Position Paper). In *Workshop on Knowledge Markup and Semantic Annotation (Semannot2004) at the International Semantic Web Conference (ISWC) 2004*, 2004. URL http://www.dfki.de/dfkibib/publications/docs/iswc04.semannot.pdf. Retrieved 2010-07-18.

P. Buitelaar and S. Ramaka. Unsupervised Ontology-based Semantic Tagging for Knowledge Markup. In *Workshop on Learning in Web Search at the International Conference on Machine Learning (ICML) 2005*, 2005. URL http://cosco.hiit.fi/search/learninginsearch05/. Retrieved 2010-10-08.

A. Bunt, C. Conati, and J. McGrenere. Supporting Interface Customization Using a Mixed-initiative Approach. In *International Conference on Intelligent User Interfaces (IUI) 2007*, pages 92–101. ACM, New York, NY, USA, 2007.

F. Buschmann, R. Meunier, H. Rohnert, P. Sommerlad, and M. Stal. *Pattern-oriented Software Architecture*. Wiley, Chichester, UK, 1996.

S. Carberry. Techniques for Plan Recognition. *User Modeling and User-Adapted Interaction*, 11(1-2):31–48, 2001.

S. Card, T. Moran, and A. Newell. *The Psychology of Human-computer Interaction*. Lawrence Erlbaum Associates, Hillsdale, NJ, USA, 1983.

F. Carmagnola, F. Cena, C. Gena, and I. Torre. A Multidimensional Framework for the Representation of Ontologies in Adaptive Hypermedia Systems. In *Advances in Artificial Intelligence (AI\*IA) 2005*, number 3673 in Lecture Notes in Artificial Intelligence, pages 370–380. Springer Verlag, Heidelberg, Germany, 2005.

S.-C. Chou, W.-T. Hsieh, F. L. Gandon, and N. M. Sadeh. Semantic Web Technologies for Context-Aware Museum Tour Guide Applications. In *International Conference on Advanced Information Networking and Applications (AINA) 2005*, pages 709–714. IEEE Computer Society, Washington, DC, USA, 2005.

J. Chu-Carroll. MIMIC: An Adaptive Mixed Initiative Spoken Dialogue System for Information Queries. In *Conference on Applied Natural Language Processing*, pages 97–104. Association for Computational Linguistics, Morristown, NJ, USA, 2000.

I. Cohen, N. Sebe, A. Garg, L. S. Chen, and T. S. Huang. Facial Expression Recognition from Video Sequences: Temporal and Static Modeling. *Computer Vision and Image Understanding, Special Issue on Face Recognition*, 91(1–2):160–187, 2003.

R. Cook and J. Kay. The Justified User Model: a Viewable, Explained User Model. In *International Conference on User Modeling (UM) 1994*, pages 145–150, User Modeling Inc., Hyannis, MA, 1994.

D. D. Corkill. Blackboard Systems. *AI Expert*, 6(9):40–47, 1991.

J. Danculovic, G. Rossi, D. Schwabe, and L. Miaton. Patterns for Personalized Web Applications. In *European Conference on Pattern Languages of Programs (EuroPLoP) 2001*, pages 423–436. Universitätsverlag Konstanz, Konstanz, Germany, 2001.

B. D. Davison and H. Hirsh. Probabilistic Online Action Prediction. Technical Report SS-98-02, AAAI Press, Menlo Park, CA, USA, 1998.

P. De Bra, P. Brusilovsky, and G.-J. Houben. Adaptive Hypermedia: From Systems to Framework. *ACM Computing Surveys*, 31(4es), 1999a. http://surveys.acm.org/, retrieved 2010-04-26.

P. De Bra, G.-J. Houben, and H. Wu. AHAM: A Dexter-based Reference Model for Adaptive Hypermedia. In *Conference on Hypertext and Hypermedia (HYPERTEXT) 1999*, pages 147–156. ACM, New York, NY, USA, 1999b.

A. Dearden and J. Finlay. Pattern Languages in HCI: A Critical Review. *Human Computer Interaction*, 21(1):49–102, 2006.

H. Dieterich, U. Malinowski, T. Kühme, and M. Schneider-Hufschmidt. State of the Art in Adaptive User Interfaces. In *Adaptive User Interfaces*, pages 13–48. North-Holland, Amsterdam, The Netherlands, 1993.

J. Dietrich and C. Elgar. A Formal Description of Design Patterns using OWL. In *Australian Software Engineering Conference*, pages 243–250. IEEE Computer Society, Los Alamitos, CA, USA, 2005.

E. W. Dijkstra. A Note on Two Problems in Connexion with Graphs. *Numerische Mathematik*, 1:269–271, 1959.

A. Dix, J. Finlay, and R. Beale. Analysis of User Behaviour as Time Series. In *Conference on People and Computers (HCI) 1992*, pages 429–444. Cambridge University Press, New York, NY, USA, 1993.

A. J. Dix, J. Finlay, and G. D. Abowd. *Human-Computer Interaction*. Pearson Prentice-Hall, Harlow, UK, 3rd edition, 2004.

P. Dolog and W. Nejdl. Using UML and XMI for Generating Adaptive Navigation Sequences in Web-Based Systems. In *Conference The Unified Modeling Language (UML) 2003*, volume 2863 of *Lecture Notes in Computer Science*, pages 205–219. Springer Verlag, Heidelberg, Germany, 2003.

P. Dolog, N. Henze, W. Nejdl, and M. Sintek. Towards the adaptive semantic web. In *Principles and Practice of Semantic Web Reasoning (PPSWR) 2003*, volume 2901 of *Lecture Notes in Computer Science*, pages 51–68. Springer Verlag, Heidelberg, Germany, 2003.

F. Dretske. *Explaining Behavior. Reasons in a World of Causes*. MIT Press, Cambridge, MA, USA, 1988.

J. Eisenstein, J. Vanderdonckt, and A. R. Puerta. Applying Model-based Techniques to the Development of UIs for Mobile Computers. In *Intelligent User Interfaces (IUI) 2001*, pages 69–76. ACM Press, New York, NY, USA, 2001.

Europäisches Kommittee für Normung. DIN 9241-110: Ergonomie der Mensch-System-Interaktion: Teil 110 Grundsätze der Dialoggestaltung, 2006.

S. Fincher, J. Finlay, S. Greene, L. Jones, P. Matchen, J. Thomas, and P. J. Molina. Perspectives on HCI Patterns: Concepts and Tools. In *Extended Abstracts on Human Factors in Computing Systems (CHI) 2003*, pages 1044–1045. ACM Press, New York, NY, USA, 2003.

L. Findlater and J. McGrenere. A Comparison of Static, Adaptive, and Adaptable Menus. In *Conference on Human Factors in Computing Systems (CHI)*
*2003*, pages 89–96. ACM Press, New York, NY, USA, 2004.

L. Findlater, K. Moffatt, J. McGrenere, and J. Dawson. Ephemeral Adaptation: The Use of Gradual Onset to Improve Menu Selection Performance.
In *Conference on Human Factors in Computing Systems (CHI) 2009*, pages
1655–1664. ACM Press, New York, NY, USA, 2009.

J. Fink, A. Kobsa, and A. Nill. User-oriented Adaptivity and Adaptability in
the AVANTI Project. In *Proceedings of Designing for the Web: Empirical*
*Studies*, Redmond, WA, USA, 1996.

D. Franklin, J. Budzik, and K. Hammond. Plan-based Interfaces: Keeping
Track of User Tasks and Acting to Cooperate. In *Conference on Intelligent*
*User Interfaces (IUI) 2002*, pages 79–86. ACM, New York, NY, USA, 2002.

K. Gajos and D. S. Weld. SUPPLE: Automatically Generating User Interfaces. In *Conference on Intelligent User Interfaces (IUI) 2004*, pages 93–
100. ACM Press, New York, NY, USA, 2004.

K. Gajos, J. O. Wobbrock, and D. S. Weld. Improving the Performance
of Motor-impaired Users with Automatically-generated, Ability-based Interfaces. In *Conference on Human Factors in Computing Systems (CHI)*
*2008*, pages 1257–1266. ACM, New York, NY, USA, 2008.

K. Z. Gajos, M. Czerwinski, D. S. Tan, and D. S. Weld. Exploring the Design
Space for Adaptive Graphical User Interfaces. In *Conference on Advanced*
*Visual Interfaces (AVI) 2006*, pages 201–208. ACM, New York, NY, USA,
2006.

U. Galassi, A. Giordana, and D. Mendola. Learning User Profile from Traces.
In *International Symposium on Applications and the Internet Workshops*
*(SAINT) 2005*, pages 166–169. IEEE Computer Society, Washington, DC,
USA, 2005.

E. Gamma, R. Helm, R. Johnson, and J. Vlissides. *Design Patterns: Elements*
*of Reusable Object-Oriented Software*. Addison-Wesley, Upper Saddle River,
NJ, USA, 1995.

A. Garland, K. Ryall, and C. Rich. Learning Hierarchical Task Models by
Defining and Refining Examples. Technical Report TR2001-26, Mitsubishi
Electric Research Laboratories, 2001.

P. Georgiakakis and S. Retalis. Design patterns for Creativity. In *European*
*Conference on Pattern Languages of Programs (EuroPLoP) 2009*. CEUR
Proceedings 566, http://CEUR-WS.org/Vol-566/, 2009.

K. Georgila and O. Lemon. Adaptive Multimodal Dialogue Management
Based on the Information State Update Approach. Technical report, University of Edinburgh, 2004.

M. Gnjatović and D. Rösner. Adaptive Dialogue Management in the
NIMITEK Prototype System. In *Perception in Multimodal Dialogue Sys-*
*tems (PIT) 2008*, volume 5078 of *Lecture Notes in Computer Science*, pages
14–25. Springer Verlag, Berlin/Heidelberg, Germany, 2008.

M. Golemati, A. Katifori, C. Vassilakis, G. Lepouras, and C. Halatsis. Creating an Ontology for the User Profile: Method and Applications. In *International Conference on Research Challenges in Information Science (RCIS) 2007*. EMSI, Casablanca, Morocco, 2007.

S. Goronzy, R. Mochales, and N. Beringer. Developing Speech Dialogs for Multimodal HMIs Using Finite State Machines. In *9th International Conference on Spoken Language Processing (Interspeech) 2006*. ICSA Archive, http://www.isca-speech.org/archive/interspeech_2006, 2006.

S. Greenberg and I. H. Witten. Adaptive Personalized Interfaces – A Question of Viability. *Behaviour and Information Technology*, 4(1):31–45, 1985.

T. R. Gruber. Toward Principles for the Design of Ontologies Used for Knowledge Sharing. *International Journal Human-Computer Studies*, 43(5–6): 907–928, 1995.

I. Gurevych, R. Porzel, E. Slinko, N. Pfleger, J. Alex, and S. Merten. Less is More: Using a Single Knowledge Representation in Dialogue Systems. In *NAACL HLT Workshop on Text Meaning*, pages 14–21. Association for Computational Linguistics, Morristown, NJ, USA, 2003.

F. Halasz and M. Schwartz. The Dexter Hypertext Reference Model. *Communications of the ACM*, 37(2):30–39, 1994.

J. O. Hallstrom and N. Soundarajan. Formalizing Design Patterns: A Comprehensive Contract for Composite. In *Workshop on the Specification and Verification of Component-Based Systems*, pages 77–82, 2008. URL http://www.eecs.ucf.edu/SAVCBS/2008/. Retrieved 2010-10-08.

H. J. Hamilton, X. Wang, and Y. Y. Yao. WebAdaptor: Designing Adaptive Web Sites Using Data Mining Techniques. In *International Florida Artificial Intelligence Research Society (FLAIRS) Conference 2001*, pages 128–132. AAAI Press, Menlo Park, CA, USA, 2001.

D. Harel. Statecharts: A Visual Formalism for Complex Systems. *Science of Computer Programming*, 8(3):231–274, 1987.

A. Harter, A. Hopper, P. Steggles, A. Ward, and P. Webster. The Anatomy of a Context-aware Application. In *International Conference on Mobile Computing and Networking (MobiCom) 1999*, pages 59–68. ACM, New York, NY, USA, 1999.

M. Hartmann and D. Schreiber. Prediction Algorithms for User Actions. In *Lernen - Wissen - Adaption (LWA) 2007*, pages 349–354. Martin-Luther-University Halle-Wittenberg, Halle, Germany, 2007.

L. Hassel and E. Hagen. Adaptation of an Automotive Dialogue System to Users Expertise and Evaluation of the System. *Computers and the Humanities*, 40(1):67–85, February 2006.

M. Hassenzahl, M. Burmester, and F. Koller. AttrakDiff: Ein Fragebogen zur Messung wahrgenommener hedonischer und pragmatischer Qualität. In *Mensch & Computer 2003*, pages 187–196. B. G. Teubner, Stuttgart/Leipzig, Germany, 2003.

D. Heckmann. Distributed User Modeling for Situated Interaction. In *35. GI Jahrestagung, Informatik 2005 – Situierung, Individualisierung und*

*Personalisierung*, volume P-67 of *Lecture Notes in Informatics*, pages 266–270. Springer Verlag, Heidelberg, Germany, 2005.

D. Heckmann, T. Schwartz, B. Brandherm, M. Schmitz, and M. von Wilamowitz-Moellendorff. Gumo – The General User Model Ontology. In *User Modeling (UM) 2005*, volume 3538 of *Lecture Notes in Artificial Intelligence*, pages 428–432. Springer Verlag, Berlin/Heidelberg, Germany, 2005.

E. O. Heierman and D. J. Cook. Improving Home Automation by Discovering Regularly Occurring Device Usage Patterns. In *International Conference on Data Mining (ICDM) 2003*, pages 537–540. IEEE Computer Society, Washington, DC, USA, 2003.

S. Henninger and P. Ashokkumar. An Ontology-Based Metamodel for Software Patterns. In *Conference on Software Engineering and Knowledge Engineering (SEKE) 2006*, pages 327–330. Knowledge Systems Institute, Skokie, IL, USA, 2006.

N. Henze, P. Dolog, and W. Nejdl. Reasoning and Ontologies for Personalized E-Learning in the Semantic Web. *Journal of Educational Technology and Society*, 7(4):82–97, 2004.

J. L. Herlocker, J. A. Konstan, and J. Riedl. Explaining collaborative filtering recommendations. In *Conference on Computer Supported Cooperative Work (CSCW) 2000*, pages 241–250. ACM, New York, NY, USA, 2000.

K. Höök. Steps to Take Before Intelligent User Interfaces Become Real. *Journal of Interacting with Computers*, 12(4):409–426, 2000.

I. Horrocks, P. P. Schneider, and F. van Harmelen. From SHIQ and RDF to OWL: The Making of a Web Ontology Language. *Journal of Web Semantics*, 1(1):7–26, 2003.

E. Horvitz. Principles of Mixed-initiative User Interfaces. In *Conference on Human Factors in Computing Systems (CHI) 1999*, pages 159–166. ACM Press, New York, NY, USA, 1999.

E. Horvitz, J. Breese, D. Heckerman, D. Hovel, and K. Rommelse. The Lumière Project: Bayesian User Modeling for Inferring the Goals and Needs of Software Users. In *Conference on Uncertainty in Artificial Intelligence (UAI) 1998*, pages 256–265, Morgan Kaufmann, San Fransisco, CA, USA, 1998.

A. Jameson. *Human-computer Interaction Handbook*, chapter Adaptive Interfaces and Agents, pages 305–330. Lawrence Erlbaum Publishers, Mahwah, NJ, USA, 1st edition, 2003.

S. Janarthanam and O. Lemon. User Simulations for Online Adaptation and Knowledge-Alignment in Troubleshooting Dialogue Systems. In *Workshop on the Semantics and Pragmatics of Dialogue (SEMDIAL) 2007*, 2008.

B. E. John. Why GOMS? *Interactions*, 2(4):80–89, 1995.

K. Jokinen, A. Kerminen, M. Kaipainen, T. Jauhiainen, G. Wilcock, M. Turunen, J. Hakulinen, J. Kuusisto, and K. Lagus. Adaptive Dialogue Systems - Interaction with Interact. In *Workshop on Discourse and Dialogue (SIGDial) 2002*, pages 64–73. Association for Computational Linguistics, Morristown, NJ, USA, 2002.

D. Kieras. *The Handbook of Task Analysis for Human-Computer Interaction*, chapter GOMS Models for Task Analysis, pages 83–116. Lawrence Erlbaum Associates, Mahwah, NJ, USA, 2003.

T. Klug and J. Kangasharju. Executable Task Models. In *International Workshop on Task Models and Diagrams (TAMODIA) 2005*, pages 119–122. ACM, New York, NY, USA, 2005.

H. Knublauch, R. W. Fergerson, N. F. Noy, and M. A. Musen. The Protégé OWL Plugin: An Open Development Environment for Semantic Web Applications. In *International Semantic Web Conference (ISWC) 2004*, volume 3298 of *Lecture Notes in Computer Science*. Springer Verlag, Berlin/Heidelberg, Germany, 2004.

A. Kobsa. Generic User Modeling Systems. *User Modeling and User-Adapted Interaction*, 11(1-2):49–63, 2001.

A. Kobsa. Privacy-enhanced Personalization. *Communications of the ACM*, 50(8):24–33, 2007.

A. Kobsa and W. Pohl. The User Modeling Shell System BGP-MS. *User Modelling and User-Adapted Interaction*, 4(2):59–106, 1995.

N. Koch and G. Rossi. Patterns for Adaptive Web Applications. In *European Conference on Pattern Languages of Programs (EuroPlop) 2002*, pages 179–194. Universitätsverlag Konstanz, Konstanz, Germany, 2002.

N. Koch and M. Wirsing. The Munich Reference Model for Adaptive Hypermedia Applications. In *International Conference on Adaptive Hypermedia and Adaptive Web-Based Systems (AH) 2002*, volume 2347 of *Lecture Notes in Computer Science*, pages 213–222. Springer Verlag, Berlin/Heidelberg, Germany, 2002.

I. Koychev and I. Schwab. Adaptation to Drifting User's Interests. In *Machine Learning in the New Information Age (ECML) Workshop 2000*, pages 39–45, 2000. URL http://www.ics.forth.gr/~potamias/mlnia.html. Retrieved 2010-10-08.

O.-W. Kwon, K. Chan, J. Hao, and T.-W. Lee. Emotion Recognition by Speech Signals. In *European Conference on Speech and Language Processing (EuroSpeech) 2003*, pages 125–128. ISCA Archive, http://www.isca-speech.org/archive/eurospeech_2003, 2003.

P. Langley. Machine Learning for Adaptive User Interfaces. In *German Conference on Artificial Intelligence (KI) 1997*, pages 53–62. Springer Verlag, Berlin/Heidelberg, Germany, 1997.

P. Langley. User Modeling in Adaptive Interfaces. In *International Conference on User Modeling (UM) 1999*, pages 357–370. Springer Verlag, New York, NY, USA, 1999.

P. Langley, C. A. Thompson, R. Elio, and A. Haddadi. An Adaptive Conversational Interface for Destination Advice. In *International Conference on Cooperative Information Agents (CIA) 1999*, pages 347–364. Springer Verlag, Berlin/Heidelberg, Germany, 1999.

S. Larsson and D. R. Traum. Information State and Dialogue Management in the TRINDI Dialogue Move Engine Toolkit. *Natural Language Engineering*, 6(3-4):323–340, 2000.

T. Lavie, J. Meyer, K. Bengler, and J. F. Coughlin. The Evaluation of In-vehicle Adaptive Systems. In *Workshop on the Evaluation of Adaptive Systems, held in conjunction with the Conference on User Modeling (UM) 2005*, pages 9–18, 2005. URL http://www.easy-hub.org/workshops/um2005/. Retrieved 2010-10-08.

O. Lemon, K. Georgila, J. Henderson, and M. Stuttle. An ISU Dialogue System Exhibiting Reinforcement Learning of Dialogue Policies: Generic Slot-filling in the TALK In-car System. In *Conference of the European Chapter of the Association for Computational Linguistics (EACL) 2006*, pages 119–122, 2006.

D. Lenat. Cyc: A Large-scale Investment in Knowledge Infrastructure. *Communications of the ACM*, 38(11):33–38, 1995.

E. Levin, R. Pieraccini, and W. Eckert. A Stochastic Model of Human-machine Interaction for Learning Dialog Strategies. *IEEE Transactions on Speech and Audio Processing*, 8(1):11–23, 2000.

Q. Limbourg, J. Vanderdonckt, B. Michotte, L. Bouillon, and V. López-Jaquero. USIXML: A Language Supporting Multi-path Development of User Interfaces. In *Engineering Human Computer Interaction and Interactive Systems (EHCI-DSVIS) 2004*, volume 3425 of *Lecture Notes in Computer Science*, pages 200–220. Springer, Heidelberg, Germany, 2004.

D. J. Litman and S. Pan. Designing and Evaluating an Adaptive Spoken Dialogue System. *User Modeling and User-Adapted Interaction*, 12(2-3):111–137, 2002.

J. Liu, C. K. Wong, and K. K. Hui. An Adaptive User Interface Based On Personalized Learning. *IEEE Intelligent Systems*, 18(2):52–57, 2003.

V. Lopez-Jaquero, J. Vanderdonckt, F. Montero, and P. Gonzalez. Towards an Extended Model of User Interface Adaptation: the ISATINE Framework. In *Engineering Interactive Systems (EIS) 2007*, volume 4940 of *Lecture Notes in Computer Science*, pages 374–392. Springer Verlag, Berlin/Heidelberg, Germany, 2007.

W. E. Mackay. Triggers and Barriers to Customizing Software. In *Conference on Human Factors in Computing Systems (CHI) 1991*, pages 153–160. ACM Press, New York, NY, USA, 1991.

P. Maes. Agents that Reduce Work and Information Overload. *Communications of the ACM*, 37(7):30–40, 1994.

H. Mannila, H. Toivonen, and A. I. Verkamo. Discovery of Frequent Episodes in Event Sequences. *Data Mining and Knowledge Discovery*, 1(3):259–289, 1997.

L. Marquardt, L. Cristoforetti, E. Mabande, N. Beringer, F. Arisio, and M. Bezold. DICIT Deliverable 6.2: Multi-microphone Data Collection and WOZ Experiments for the Analysis of User Behaviour in the DICIT Scenarios. Public deliverable, Fondazione Bruno Kesseler, Trento, Italy, 2008.

M. Matassoni, M. Omologo, R. Manione, T. Sowa, R. Balchandran, M. E. Epstein, and L. Seredi. The DICIT Project: An Example of Distant-talking Based Spoken Dialogue Interactive System. In *Intelligent Information Systems (IIS) 2008*, pages 527–533. Academic Publishing House EXIT, Warsaw, Poland, 2008.

M. F. McTear. *Spoken Dialogue Technology*. Springer Verlag, London, UK, 2004.

E. Metzker, A. Seffah, and A. Gaffar. Towards a Systematic Empirical Validation of HCI Knowledge Captured as Patterns. In *International Conference on Human-Computer Interaction (HCI) 2003*, pages 168–172. Lawrence Erlbaum Publishers, Mahwah, NJ, USA, 2003.

T. Mikkonen. Formalizing Design Patterns. In *International Conference on Software engineering (ICSE) 1998*, pages 115–124. IEEE Computer Society, Washington, DC, USA, 1998.

D. Milward and M. Beveridge. Ontology-based Dialogue Systems. In *IJCAI 2003 Workshop on Knowledge and Reasoning in Practical Dialogue Systems*, pages 9–18, 2003. URL http://www.ida.liu.se/~nlplab/ijcai-ws-03/. Retrieved 2010-10-08.

J. Mitchell and B. Shneiderman. Dynamic versus Static Menus: an Exploratory Comparison. *SIGCHI Bulletins*, 20(4):33–37, 1989.

T. M. Mitchell. *Machine Learning*. McGraw Hill, Singapore, 1997.

N. Mitrović, J. A. Royo, and E. Mena. Performance Analysis of an Adaptive User Interface System Based on Mobile Agents. In *Engineering Interactive Systems (EIS) 2008*, volume 4940 of *Lecture Notes in Computer Science*, pages 1–17. Springer Verlag, Berlin/Heidelberg, Germany, 2008.

C. H. Mooney. *The Discovery of Interacting Episodes and Temporal Rule Determination in Sequential Pattern Mining*. PhD thesis, Flinders University, Adelaide, Australia, 2007.

G. Mori, F. Paternò, and C. Santoro. CTTE: Support for Developing and Analyzing Task Models for Interactive System Design. *IEEE Transactions on Software Engineering*, 28(8):797–813, 2002.

G. Niklfeld, R. Finan, and M. Pucher. Architecture for Adaptive Multimodal Dialog Systems Based on VoiceXML. In *European Conference on Speech and Language Processing (EuroSpeech) 2001*. ISCA Archive, http://www.isca-speech.org/archive/eurospeech_2001, 2001.

D. A. Norman. *User Centered System Design*, chapter Cognitive Engineering, pages 31–61. Lawrence Erlbaum Associates, Hillsdale, NJ, USA, 1986.

N. F. Noy and D. L. McGuinness. Ontology Development 101: A Guide to Creating Your First Ontology. Technical Report KSL-01-05, Stanford Knowledge Systems Laboratory, March 2001.

N. F. Noy and M. A. Musen. PROMPT: Algorithm and Tool for Automated Ontology Merging and Alignment. In *National Conference on Artificial Intelligence and Conference on on Innovative Applications of Artificial Intelligence (AAAI/IAAI) 2000*, pages 450–455. AAAI Press, Menlo Park, CA, USA, 2000.

D. Oberle, A. Ankolekar, P. Hitzler, P. Cimiano, M. Sintek, M. Kiesel, B. Mougouie, S. Baumann, S. Vembu, M. Romanelli, P. Buitelaar, R. Engel, D. Sonntag, N. Reithinger, B. Loos, H.-P. Zorn, V. Micelli, R. Porzel, C. Schmidt, M. Weiten, F. Burkhardt, and J. Zhou. DOLCE ergo SUMO: On Foundational and Domain Models in SWIntO (SmartWeb Integrated Ontology). *Journal of Web Semantics: Science, Services and Agents on the World Wide Web*, 5(3):156–174, 2007.

Z. Obrenović, D. Starĉević, and V. Devedẑić. Using Ontologies in the Design of Multimodal User Interfaces. In *International Conference on Human-Computer Interaction (Interact) 2003*, pages 535–542. IOS Press, Amsterdam, The Netherlands, 2003.

R. Oppermann. Adaptively Supported Adaptability. *International Journal of Human Computer Studies*, 40(3):455–472, 1994.

J. L. Ortega-Arjona. Applying Architectural Patterns for Parallel Programming. Solving the One-dimensional Heat Equation. In *European Conference on Pattern Languages of Programs (EuroPLoP) 2009*. CEUR Proceedings 566, http://CEUR-WS.org/Vol-566/, 2009.

J. Orwant. Heterogeneous Learning in the Doppelgänger User Modeling System. *User Modeling and User-Adapted Interaction*, 4(2):107–130, 1995.

S. Oviatt. Ten Myths of Multimodal Interaction. *Communications of the ACM*, 42(11):74–81, 1999.

T. Paek. Reinforcement Learning for Spoken Dialogue Systems: Comparing Strengths and Weaknesses for Practical Deployment. Technical Report MSR-TR-2006-62, Microsoft Corp., 2006.

A. Paramythis and S. Weibelzahl. A Decomposition Model for the Layered Evaluation of Interactive Adaptive Systems. In *User Modeling (UM) 2005*, volume 3538 of *Lecture Notes in Computer Science*, pages 438–442. Springer Verlag, Berlin/Heidelberg, Germany, 2005.

A. Paramythis, A. Totter, and C. Stephanidis. A Modular Approach to the Evaluation of Adaptive User Interfaces. In *Workshop on the Empirical Evaluation of Adaptive Systems, held at User Model (UM) 2001*, pages 9–24, 2001. URL http://www.easy-hub.org/workshops/um2001/. Retrieved 2010-10-08.

J. Park, S. H. Han, Y. S. Park, and Y. Cho. Usability of Adaptable and Adaptive Menus. In *Usability and Internationalization. HCI and Culture*, volume 4559 of *Lecture Notes in Computer Science*, pages 405–411. Springer Verlag, Berlin/Heidelberg, Germany, 2007.

F. Paternò. *Handbook of Software Engineering and Knowledge Engineering*, chapter Task Models in Interactive Software Systems, pages 817–836. World Scientific Publishing Company, Singapore, 2001.

F. Paternò, C. Mancini, and S. Meniconi. ConcurTaskTrees: A Diagrammatic Notation for Specifying Task Models. In *International Conference on Human-Computer Interaction (Interact) 1997*, pages 362–369. Chapman & Hall, London, UK, 1997.

M. Perkowitz and O. Etzioni. Adaptive Web Sites: an AI Challenge. In *International Joint Conference on Artifical Intelligence (IJCAI) 1997*, pages 16–23. Morgan Kaufmann, San Francisco, CA, USA, 1997.

M. Perkowitz and O. Etzioni. Adaptive Web Sites: Conceptual Cluster Mining. In *International Joint Conference on Artifical Intelligence (IJCAI) 1999*, pages 264–269. Morgan Kaufmann, San Francisco, CA, USA, 1999.

R. Petrasch. Model Based User Interface Design: Model Driven Architecture und HCI Patterns. *GI Softwaretechnik-Trends*, 27(3):5–10, 2007.

J. Pittermann, W. Minker, A. Pittermann, and D. Bühler. PROBLEMO - Problem Solving and Emotion Awareness in Spoken Dialogue Systems. In *International Conference on Intelligent Environments (IE) 2007*, pages 447–450. IEEE Computer Society, Washington, DC, USA, 2007.

R. Porzel, N. Pflege, S. Merten, M. Löckelt, I. Gurevych, R. Engel, and J. Alexandersson. More on Less: Further Applications of Ontologies in Multi-Modal Dialogue Systems. In *IJCAI Workshop on Knowledge and Reasoning in Practical Dialogue Systems*, 2003. URL http://www.ida.liu.se/~arnjo/Ijcai09ws/. Retrieved 2010-10-08.

M. Prensky. Digital Natives, Digital Immigrants. *On the Horizon*, 9(5), 2001.

E. Prud'hommeaux and A. Seaborne. SPARQL Query Language for RDF. W3C recommendation, World Wide Web Consortium, 2008. http://www.w3.org/TR/rdf-sparql-query/, retrieved 2010-05-26.

L. Razmerita, A. A. Angehrn, and A. Maedche. Ontology-Based User Modeling for Knowledge Management Systems. In *User Modeling (UM) 2003*, volume 2702 of *Lecture Notes in Computer Science*, pages 213–217. Springer Verlag, Berlin/Heidelberg, Germany, 2003.

K. Reinecke and A. Bernstein. Tell Me Where You've Lived, and I'll Tell You What You Like: Adapting Interfaces to Cultural Preferences. In *International Conference on User Modeling, Adaptation, and Personalization (UMAP) 2009*, volume 5535 of *Lecture Notes in Computer Science*, pages 185–196. Springer Verlag, Berlin/Heidelberg, Germany, 2009.

E. Rich. User Modeling via Stereotypes. *Cognitive Science: A Multidisciplinary Journal*, 3(4):329–354, 1979.

E. Rich. Users are Individuals: Individualizing User Models. *International Journal of Man-Machine Studies*, 18(3):199–214, 1983.

V. Rieser and O. Lemon. Learning Effective Multimodal Dialogue Strategies from Wizard-of-Oz Data: Bootstrapping and Evaluation. In *Annual Meeting of the Association for Computational Linguistics (ACL/HLT) 2008*, pages 638–646. Association for Computational Linguistics, Morristown, NJ, USA, 2008.

V. Rieser and O. Lemon. Does This List Contain What You Were Searching for? Learning Adaptive Dialogue Strategies for Interactive Question Answering. *Natural Language Engineering*, 15(1):55–72, 2009.

R. R. Sarukkai. Link Prediction and Path Analysis Using Markov Chains. In *International World Wide Web Conference on Computer Networks*, pages 377–386. North-Holland Publishing, Amsterdam, The Netherlands, 2000.

K. Scheffler and S. Young. Automatic Learning of Dialogue Strategy Using Dialogue Simulation and Reinforcement Learning. In *International Conference on Human Language Technology Research*, pages 12–19. Morgan Kaufmann Publishers, San Francisco, CA, USA, 2002.

C. F. Schmidt, N. S. Sridharan, and J. L. Goodson. The Plan Recognition Problem. *Artificial Intelligence*, 11:45–83, 1978.

D. C. Schmidt. Model-Driven Engineering. *IEEE Computer*, 39(2):25–31, 2006.

K.-U. Schmidt, J. Dörflinger, T. Rahmani, M. Sahbi, L. Stojanovic, and S. M. Thomas. An User Interface Adaptation Architecture for Rich Internet Applications. In *The Semantic Web: Research and Applications*, volume 5021 of *Lecture Notes in Computer Science*, pages 736–750. Springer Verlag, Berlin/Heidelberg, Germany, 2008.

N. Schneider, S. Schreiber, J. Wilkes, M. Grandt, and C. M. Schlick. Investigation of Adaptation Dimensions for Age-Differentiated Human-Computer Interfaces. In *Universal Acess in Human Computer Interaction. Coping with Diversity*, volume 4554 of *Lecture Notes in Computer Science*, pages 1010–1019. Springer Verlag, Berlin/Heidelberg, Germany, 2007.

A. Sears and B. Shneiderman. Split Menus: Effectively Using Selection Frequency to Organize Menus. *ACM Transactions on Computer-Human Interaction*, 1(1):27–51, 1994.

G. Sharifi, R. Deters, J. Vassileva, S. Bull, and H. Röbig. Location-Aware Adaptive Interfaces for Information Access with Handheld Computers. In *Adaptive Hypermedia and Adaptive Web-Based Systems*, volume 3137 of *Lecture Note in Computer Science*, pages 305–328. Springer, Berlin/Heidelberg, Germany, 2004.

B. Shneiderman and P. Maes. Direct Manipulation vs. Interface Agents. *Interactions*, 4:42–61, 1997.

B. Shneiderman and C. Plaisant. *Designing the User Interface: Strategies for Effective Human-Computer Interaction*. Addison Wesley, Reading, MA, USA, 4th edition, April 2004.

S. Singh, D. Litman, M. Kearns, and M. Walker. Optimizing Dialogue Management with Reinforcement Learning: Experiments with the NJFun System. *Journal of Artifical Intelligence Research*, 16(1):105–133, 2002.

M. K. Smith, C. Welty, and D. L. McGuinness. OWL Web Ontology Language Guide. W3C recommendation, World Wide Web Consortium, 2004. http://www.w3.org/TR/owl-guide/, retrieved 2010-04-26.

D. Sonntag, R. Engel, G. Herzog, A. Pfalzgraf, N. Pfleger, M. Romanelli, and N. Reithinger. SmartWeb Handheld - Multimodal Interaction with Ontological Knowledge Bases and Semantic Web Services. In *Artifical Intelligence for Human Computing*, number 4451 in Lecture Notes in Artificial Intelligence, pages 272–295. Springer, Berlin/Heidelberg, Germany, 2007.

T. Sowa and F. Arisio. DICIT Deliverable 6.5: Final STB Prototype Evaluation. Public deliverable, Fondazione Bruno Kesseler, Trento, Italy, 2009.

C. Stephanidis, A. Paramythis, C. Karagiannidis, and A. Savidis. Supporting Interface Adaptation in the AVANTI Web Browser. In *3rd ERCIM Workshop on User Interfaces for All*, 1997.

C. Stephanidis, A. Paramythis, D. Akoumianakis, and M. Sfyrakis. Self-adapting web-based systems: Towards universal accessibility. In *4th ERCIM Workshop on User Interfaces for All*, 1998.

L. Swartz. Why People Hate the Paperclip: Labels, Appearance, Behavior, and Social Responses to User Interface Agents. Master's thesis, Symbolic Systems Program, Stanford University, 2003.

D. V. Thompson, R. W. Hamilton, and R. Rust. Feature Fatigue: When Product Capabilities Become Too Much of a Good Thing. *Journal of Marketing Research*, 42:431–442, 2005.

J. Tidwell. *Designing Interfaces*. O'Reilly Media, Sebastopol, CA, USA, 2005.

T. Tran, P. Cimiano, and A. Ankolekar. A Rule-Based Adaption Model for Ontology-Based Personalization. In *Advances in Semantic Media Adaptation and Personalization*, volume 93 of *Studies in Computational Intelligence*, pages 117–135. Springer Verlag, Berlin/Heidelberg, Germany, 2008.

T. Tsandilas and m. schraefel. An Empirical Assessment of Adaptation Techniques. In *Extended Abstracts on Human Factors in Computing Systems (CHI) 2005*, pages 2009–2012. ACM, New York, NY, USA, 2005.

D. K. van Duyne, J. A. Landay, and J. I. Hong. *The Design of Sites: Patterns for Creating Winning Web Sites*. Prentice Hall, Upper Saddle River, NJ, USA, 2nd edition, 2006.

M. van Welie and G. C. van der Veer. Pattern Languages in Interaction Design: Structure and Organization. In *International Conference on Human-Computer Interaction (Interact) 2003*, pages 527–534. IOS Press, Amsterdam, The Netherlands, 2003.

G. Veldhuijzen van Zanten. Adaptive Mixed-initiative Dialogue Management. In *Interactive Voice Technology for Telecommunications Applications (IVTTA) 1998*, pages 65–70. IEEE Computer Society, Washington, DC, USA, 1998.

G. Veldhuijzen van Zanten. User Modelling in Adaptive Dialogue Management. In *European Conference on Speech Communication and Technology (EuroSpeech) 1999*, pages 1183–1186. ISCA Archive, http://www.isca-speech.org/archive/eurospeech_1999, 1999.

E. Vidal, F. Thollard, C. de la Higuera, F. Casacuberta, and R. C. Carrasco. Probabilistic Finite-State Machines-Part I. *IEEE Transactions on Pattern Analysis and Machine Intelligence*, 27(7):1013–1025, 2005a.

E. Vidal, F. Thollard, C. de la Higuera, F. Casacuberta, and R. C. Carrasco. Probabilistic Finite-State Machines-Part II. *IEEE Transactions on Pattern Analysis and Machine Intelligence*, 27(7):1026–1039, 2005b.

E. Vildjiounaite, V. Kyllönen, T. Hannula, and P. Alahuhta. Unobtrusive Dynamic Modelling of TV Program Preferences in a Household. In *Changing Television Environments (EUROITV) 2008*, volume 5066 of *Lecture Notes in Computer Science*, pages 82–91. Springer, Berlin/Heidelberg, Germany, 2008.

W. Wahlster. SmartKom: Symmetric Multimodality in an Adaptive and Reusable Dialogue Shell. In *Human Computer Interaction Status Conference*, pages 47–62, 2003.

W. Wahlster. *40 Jahre Informatikforschung in Deutschland*, chapter SmartWeb - Ein multimodales Dialogsystem für das semantische Web. Springer Verlag, Berlin/Heidelberg, Germany, 2007.

G. I. Webb, M. J. Pazzani, and D. Billsus. Machine Learning for User Modeling. *User Modeling and User-Adapted Interaction*, 11(1-2):19–29, 2001.

G. Weber and P. Brusilovsky. ELM-ART: An Adaptive Versatile System for Web-based Instruction. *International Journal of Artificial Intelligence in Education*, 12:351–384, 2001.

D. S. Weld, C. Anderson, P. Domingos, O. Etzioni, K. Gajos, T. Lau, and S. Wolfman. Automatically Personalizing User Interfaces. In *International Joint Conference on Artificial Intelligence (IJCAI) 2003*, pages 1613–1619. Morgan Kaufmann Publishers, San Francisco, CA, USA, 2003.

H. Wesseling, M. Bezold, and N. Beringer. Automatic Evaluation Tool for Multimodal Dialogue Systems. In *Perception and Interactive Technologies for Speech-Based Systems (PIT) 2008*, volume 5078 of *Lecture Notes in Computer Science*, pages 297–305. Springer, Berlin/Heidelberg, Germany, 2008.

I. H. Witten and E. Frank. *Data Mining*. Morgan Kaufman, San Francisco, CA, USA, 2nd edition, 2005.

S. Young, M. Gaica, S. Keizera, F. Mairessea, J. Schatzmanna, B. Thomsona, and K. Yua. The Hidden Information State Model: A Practical Framework for POMDP-based Spoken Dialogue Management. *Computer, Speech and Language*, 24(2):150–174, 2010.

C. Zannier and F. Maurer. Tool Support for Complex Refactoring to Design Patterns. In *Extreme Programming and Agile Processes in Software Engineering (XP)*, volume 2675 of *Lecture Notes in Computer Science*, pages 123–130. Springer, Berlin/Heidelberg, Germany, 2003.

U. Zdun. Systematic Pattern Selection Using Pattern Language Grammars and Design Space Analysis. *Software - Practice and Experience*, 37:983–1016, 2007.

J. Zhu, J. Hong, and J. G. Hughes. Using Markov Chains for Link Prediction in Adaptive Web Sites. In *Soft-Ware 2002: Computing in an Imperfect World*, volume 2311 of *Lecture Notes in Computer Science*, pages 55–66. Springer Verlag, Berlin/Heidelberg, Germany, 2002.

I. Zukerman and D. W. Albrecht. Predictive Statistical Models for User Modeling. *User Modeling and User-Adapted Interaction*, 11(1-2):5–18, 2001.

# Index